The Rise and Fall
of National Test Scores

D1262499

EDUCATIONAL PSYCHOLOGY

Allen J. Edwards, Series Editor
Department of Psychology
Southwest Missouri State University
Springfield, Missouri

In preparation
Judith Worell (ed.). Psychological Development in the Elementary Years
Published
John B. Biggs and Kevin F. Collis. Evaluating the Quality of Learning: The Solo Taxonomy (Structure of the Observed Learning Outcome)
Gilbert R. Austin and Herbert Garber (eds.). The Rise and Fall of National Test Scores
Lynne Feagans and Dale C. Farran (eds.). The Language of Children Reared in Poverty: Implications for Evaluation and Intervention
Patricia A. Schmuck, W. W. Charters, Jr., and Richard O. Carlson (eds.). Educational Policy and Management: Sex Differentials
Phillip S. Strain and Mary Margaret Kerr. Mainstreaming of Children in Schools: Research and Programmatic Issues
Maureen L-Pope and Terence R. Keen. Personal Construct Psychology and Education
Ronald W. Henderson (ed.). Parent–Child Interaction: Theory, Research, and Prospects
W. Ray Rhine (ed.). Making Schools More Effective: New Directions from Follow Through
Herbert J. Klausmeier and Thomas S. Sipple. Learning and Teaching Concepts: A Strategy for Testing Applications of Theory
James H. McMillan (ed.). The Social Psychology of School Learning
M. C. Wittrock (ed.). The Brain and Psychology
Marvin J. Fine (ed.). Handbook on Parent Education
Dale G. Range, James R. Layton, and Darrell L. Roubinek (eds.). Aspects of Early Childhood Education: Theory to Research to Practice
Jean Stockard, Patricia A. Schmuck, Ken Kempner, Peg Williams, Sakre K. Edson, and Mary Ann Smith. Sex Equity in Education
James R. Layton. The Psychology of Learning to Read
Thomas E. Jordan. Development in the Preschool Years: Birth to Age Five
Gary D. Phye and Daniel J. Reschly (eds.). School Psychology: Perspectives and Issues

The list of titles in this series continues on page xviii of this volume.

The Rise and Fall
of National Test Scores

EDITED BY

Gilbert R. Austin

Herbert Garber

Center for Educational Research and Development
University of Maryland Baltimore County
Catonsville, Maryland

 1982

ACADEMIC PRESS
A Subsidiary of Harcourt Brace Jovanovich, Publishers
New York   London
Paris   San Diego   San Francisco   São Paulo   Sydney   Tokyo   Toronto

To the Children

COPYRIGHT © 1982, BY ACADEMIC PRESS, INC.
ALL RIGHTS RESERVED.
NO PART OF THIS PUBLICATION MAY BE REPRODUCED OR
TRANSMITTED IN ANY FORM OR BY ANY MEANS, ELECTRONIC
OR MECHANICAL, INCLUDING PHOTOCOPY, RECORDING, OR ANY
INFORMATION STORAGE AND RETRIEVAL SYSTEM, WITHOUT
PERMISSION IN WRITING FROM THE PUBLISHER.

ACADEMIC PRESS, INC.
111 Fifth Avenue, New York, New York 10003

United Kingdom Edition published by
ACADEMIC PRESS, INC. (LONDON) LTD.
24/28 Oval Road, London NW1 7DX

Library of Congress Cataloging in Publication Data
Main entry under title:

The Rise and fall of national test scores.

  (Educational psychology)
  Includes bibliographies and index.
  Contents: Learning, schooling, scores / Gilbert
Austin, Herbert Garber -- College entrance examination
trends / Bruce K. Eckland -- Analyzing changes in school
levels of achievement for men and women using Project
Talent ten-and fifteen-year retests / John C. Flanagan
-- [etc.]
  1. Educational tests and measurements--United States--
Evaluation--Addresses, essays, lectures. 2. Academic
achievement--Evaluation--Addresses, essays, lectures.
I. Austin, Gilbert R. II. Garber, Herbert. III. Series.
LB3051.R56   371.2'6   81-17668
ISBN 0-12-068580-9   AACR2

PRINTED IN THE UNITED STATES OF AMERICA

82 83 84 85   9 8 7 6 5 4 3 2 1

Contents

3

Analyzing Changes in School Levels of Achievement for Men and Women Using Project TALENT Ten- and Fifteen-Year Retests

John C. Flanagan

4

Functional Literacy and Writing: Some Cautions about Interpretation

Roy H. Forbes

5

Reading Trend Data in the United States: A Mandate for Caveats and Caution

Roger Farr and Leo Fay

6

Trends in School Mathematics Performance

James T. Fey and Thomas Sonnabend

7

**National Science Test Scores:
Positive and Negative Trends in Academic Achievement
in Science with Particular Emphasis
on the Effects of Recent Curriculum Revision**   **163**

Henry H. Walbesser and Cheryl Gonce-Winder

8

Trends in Educational Standards in Great Britain and Ireland   **195**

Thomas Kellaghan and George F. Madaus

9

Race, Social Class, Expection, and Achievement   **215**

David T. Harris and Joseph L. Amprey, Jr.

10

**The Final Hurdle: Minimum Competency
Achievement Testing**   **223**

Richard M. Jaeger

11

The Implications for Society   247

Gilbert R. Austin and Herbert Garber

Index   257

List of Contributors

Numbers in parentheses indicate the pages on which the authors' contributions begin.

Joseph L. Amprey, Jr. (215), Counseling Center, Coppin State College, Baltimore, Maryland 21216

Gilbert R. Austin (1, 247), Center for Educational Research and Development, University of Maryland Baltimore County, Catonsville, Maryland 21228

Robert L. Ebel (xiii), College of Education, Michigan State University, East Lansing, Michigan 48824

Bruce K. Eckland (9), Department of Sociology, University of North Carolina at Chapel Hill, Chapel Hill, North Carolina 27514

Roger Farr (83), School of Education, Indiana University, Bloomington, Indiana 47401

Leo Fay (83), School of Education, Indiana University, Bloomington, Indiana 47401

James T. Fey (143), College of Education, University of Maryland, College Park, Maryland 20742

John C. Flanagan (35), American Institutes for Research, Palo Alto, California 94302

Roy H. Forbes (51), National Assessment of Educational Progress, Denver, Colorado 80295

Herbert Garber (1, 247), Center for Educational Research and Development, University of Maryland Baltimore County, Catonsville, Maryland 21228

David T. Harris* (215), Office of Equal Opportunity Recruitment, University of Maryland, College Park, Maryland 20742

Richard M. Jaeger (223), School of Education, University of North Carolina at Greensboro, Greensboro, North Carolina 27412

Thomas Kellaghan (195), Educational Research Centre, St. Patrick's College, Dublin 9, Ireland

George F. Madaus (195), College of Education, Boston College, Chestnut Hill, Massachusetts 02167

Thomas Sonnabend (143), College of Education, University of Maryland, College Park, Maryland 20742

Henry H. Walbesser (163), Department of Education, University of Maryland Baltimore County, Catonsville, Maryland 21228

Cheryl Gonce-Winder (163), Department of Education, University of Maryland Baltimore County, Catonsville, Maryland 21228

* *Present address:* Lawton, Oklahoma 73503

Foreword

During the 1970s, reports of substantial declines in scores on college admission tests attracted much attention, and aroused considerable concern. Were the reports true? What did they mean? Why had it happened? Rightly or wrongly, the test score declines were often accepted as evidence of deterioration in the quality of public education in the United States.

It is not quite correct, however, to say that reports of declining test scores "aroused" the concern. The concern was already there, and had been there for about as long as schools have existed. The goals of education, and the extent to which those goals are being achieved, are matters of perennial concern to parents and to public-spirited citizens generally.

In the contemporary educational enterprise, two sets of values and two systems of belief compete for preference and influence. Essentially, the same competition has been going on for many centuries, with one set of views or the other temporarily in ascendancy. The two competitors may be designated roughly as culturalism (Aristotle) versus naturalism (Rousseau), subject-centered versus child-centered education, essentialism (Bagley) versus progressivism (Dewey), achievement versus adjustment, excellence versus equality, classical humanism versus contemporary humanism.

During periods of threat to the nation, excellence, achievement, and essentialism tend to express the will of the people. When the threat has passed, the naturalistic "let's enjoy life" philosophy tends to prevail. The peaceful decades between the two World Wars were the heyday of child-centered, progressive education. The outbreak of World War II sent teachers and pupils back to the books to hard study of math and science, languages and history. With peace came a relaxation of stress on learning and increased concern for adjustment to living. Then the Russian success in the satellite race refocused attention on excellence in learning. As that race was being won, and with the economy flourishing, concern shifted from excellence to equality, not just equality of opportunity but equality of outcomes. Collective action to achieve social goals took priority in the minds of many college students over individual efforts to acquire knowledge. But by 1981 with jobs scarce and hard times imminent, the pendulum was swinging back toward individual efforts to achieve.

While the public emphasis may shift with shifting circumstances, individual dispositions and biases tend to be more consistent. Some are consistently concerned with the pupil's intellectual development. Others are consistently concerned with the pupil's happiness and adjustment. Few are inclined to value one exclusively and to ignore the other completely, but there are sharp differences among human beings in the amount of concern they have for learning and for feelings.

These differences were reflected in the reactions of different persons to news of the declining test scores. Those who placed high value on excellence in learning almost welcomed the news of declining scores as evidence in support of the need for schools to place more emphasis on academic achievement. Those more humanistically inclined did not welcome the news of declining scores, for it seemed to point to a shortcoming of the values and methods they espoused. Perhaps, they suggested, the declines were due to changes in the tests, to changes in the composition of the group taking the tests, to changes in the importance of the scores, and so on.

Neither those who placed value on achievement nor those who placed highest value on adjustment were eager to admit that placing higher value on one inevitably resulted in less on the other. Both achievement and adjustment are good. Most of us would like more of each. But they compete for the same time and energy. The more pupils study the less time they have to play.

What is true of schools is true of our individual lives and of the welfare of our society. Within the constraints of our own talents and circumstances we try to maximize happiness by balancing the drives to achieve against the wish to enjoy. Within the constraints set by the problems it faces and the resources at its disposal, a society seeks to protect its people and contribute to their happiness. How we balance the competing demands of work and play has much to do with our individual welfare, and with the welfare of our society.

So at this moment in history, one of the important questions facing the schools in our society is, as it has often been in the past, "How much emphasis shall we place on a pursuit of excellence in learning that is certain to cost hard work and some sacrifice of immediate satisfactions?" A study of the nature and causes of the decline in test scores will not answer that question decisively, but it can contribute to the decision. That, in my opinion, is the ultimate value of this volume.

Professor Austin and Professor Garber have enlisted the help of a number of distinguished scholars to comment on various aspects of the score declines. Using data from a variety of sources, they discuss educational trends during the last 50 years; possible causes of the score declines; trends in various areas of study and in some other countries; black concerns with aptitude testing; and the recent movement toward testing of minimum competency. This broad survey of current developments in education will elaborate and refine our understanding of the score declines. It should help us to decide what to do about the situation in which we find ourselves. Professors Austin and Garber and the contributors to this volume deserve our thanks and congratulations for a job well done.

Robert L. Ebel
MICHIGAN STATE UNIVERSITY

Preface

Every society values education. The institutions it creates for instruction are considered essential to the very survival of the society itself. When evidence emerges suggesting that the outcome of study in the educational institutions is less than it ought to be, public concern is aroused. Test scores are used to measure the health of an educational process. When declining test scores in the nation's schools are reported, public concern is understandably aroused.

This book examines, in some depth, the nature of test score changes over an extended period of time and in a broad range of subject matters and levels of schooling. It aims to enable educators, researchers, political leaders, and informed, interested laymen to be able to understand the complexities underlying simplified periodic publication of test score trends and momentary changes in test score averages.

The volume contains chapters contributed by experts on various aspects of educational tests and score interpretation and uses. The reader will obtain insights into interpretations of trends in college admissions test scores and questions about the postadmission performance of ethnic minority students (Eckland, Chapter 2, and Harris and Amprey, Chapter 9). Separate chapters are devoted to an examination of general school achievement trends of high school students (Flanagan, Chapter 3) and the trends observed in broad-based testing programs of the National Assessment of Educational Progress (NAEP) (Forbes, Chapter 4).

The complexities involved in obtaining dependable data with which to make informed judgments about reading achievement trends are examined and discussed in depth by Farr and Fay, Chapter 5. The applicability to all subject matter areas of the nearly insurmountable difficulties will be apparent to the reader.

Attention is given in Chapter 7 by Walbesser and Gonce-Winder to the crucial problem of adequate test-item construction through an analysis of NAEP science items used in recent national assessments of 9-, 13-, and 17-year-old students. Here, too, findings will probably strike the reader as applicable to other subject matter areas. Chapter 6 by Fey and Sonnabend deals with problems of mathematics testing and focuses on standardizing testing procedures and the extent to which inadequate control can lead to invalid test scores.

To give a broader perspective to the problem, Kellaghan and Madaus (Chapter 8) describe and discuss test score trends and problems in Great Britain and Ireland, with attention given to the similarities and differences of those countries and the United States. For further breadth, the chapter deals at length with attempts to improve school achievement problems through the minimum competency testing movement. The reader will be reminded once again, in a striking manner, that complex problems seldom yield satisfactory solutions from simple policy changes.

This book addresses the general problem from the perspectives of such disciplines as sociology, policy and political science, and psychology as well as general education, educational administration and tests and measurements. To the knowledge of the editors, no other publication is presently in print that takes so broad an approach to the specific problem of changing test scores.

Garber and Austin in Chapter 1 place the problem in context and describe the guidelines each contributor was asked to use in writing his or her chapter. Chapter 11, by Austin and Garber, summarizes and reviews the contributed chapters and attempts to identify and sharpen the several themes which emerge and to provide a statement of implication and meaning.

The editors acknowledge, gratefully, the help and support of the University of Maryland Baltimore County and many individuals. Special thanks are extended to Lawrence Stolorow of the University of Iowa for his early association with this publication that led to the overall plan for the book and to Richard Neville, Joseph Mulligan, and William Johnson of the University of Maryland Baltimore County for their encouragement and support. The editors wish to express their gratitude to Ronald Zeigler for early help with the volume. The editors are also indebted to the American Educational Research Association and the National Council on Measurement in Education for the joint symposium held in 1977 to discuss the rise and fall of test scores, which is where the idea for this book was conceived. Due acknowledgment and thanks are extended to Shirley Alonso, Audrey Mahoney, Eleanor Austin, and Doris Garber for the many patient hours spent in text and table preparations. Finally, many thanks are due the chapter authors for the great care and enduring good cheer with which they reviewed their work as requested by the editors.

EDUCATIONAL PSYCHOLOGY

continued from page ii

Norman Steinaker and M. Robert Bell. The Experiential Taxonomy: A New Approach to Teaching and Learning

J. P. Das, John R. Kirby, and Ronald F. Jarman. Simultaneous and Successive Cognitive Processes

Herbert J. Klausmeier and Patricia S. Allen. Cognitive Development of Children and Youth: A Longitudinal Study

Victor M. Agruso, Jr. Learning in the Later Years: Principles of Educational Gerontology

Thomas R. Kratochwill (ed.). Single Subject Research: Strategies for Evaluating Change

Kay Pomerance Torshen. The Mastery Approach to Competency-Based Education

Harvey Lesser. Television and the Preschool Child: A Psychological Theory of Instruction and Curriculum Development

Donald J. Treffinger, J. Kent Davis, and Richard E. Ripple (eds.). Handbook on Teaching Educational Psychology

Harry L. Hom, Jr. and Paul A. Robinson (eds.). Psychological Processes in Early Education

J. Nina Lieberman. Playfulness: Its Relationship to Imagination and Creativity

Samuel Ball (ed.). Motivation in Education

Erness Bright Brody and Nathan Brody. Intelligence: Nature, Determinants, and Consequences

António Simões (ed.). The Bilingual Child: Research and Analysis of Existing Educational Themes

Gilbert R. Austin. Early Childhood Education: An International Perspective

Vernon L. Allen (ed.). Children as Teachers: Theory and Research on Tutoring

Joel R. Levin and Vernon L. Allen (eds.). Cognitive Learning in Children: Theories and Strategies

Donald E. P. Smith and others. A Technology of Reading and Writing (in four volumes).

> Vol. 1. *Learning to Read and Write: A Task Analysis (by Donald E. P. Smith)*
> Vol. 2. *Criterion-Referenced Tests for Reading and Writing (by Judith M. Smith, Donald E. P. Smith, and James R. Brink)*
> Vol. 3. *The Adaptive Classroom (by Donald E. P. Smith)*
> Vol. 4. *Designing Instructional Tasks (by Judith M. Smith)*

Phillip S. Strain, Thomas P. Cooke, and Tony Apolloni. Teaching Exceptional Children: Assessing and Modifying Social Behavior

Chapter 1

Learning, Schooling, Scores: A Continuing Controversy

Herbert Garber
Gilbert R. Austin

There has been, in recent times, a growing concern across America about declining test scores. Over the past 15 years, media stories have convinced us that students in American schools are uniformly learning less of what they need to know to assume competent adult roles. The fact, however, is that standardized test scores have been *both* falling and rising. It is a tired truism that no news is good news, but too often the corollary, good news is no news, is the case. This may explain, in part, why most Americans have not heard about rising test scores. And the media, of course, have found that as public education consumes a larger share of national and local resources, it requires examination and criticism. That is as it should be. However, the news media are not principals in the debate; they are recorders and couriers. This book contains the best thinking of concerned professionals.

The public, many of whom are parents of school-age children, feel justifiable concern. This concern has been expressed in many ways. One most significant outcome is political–legal action: minimum competency achievement legislation, test and test-item score disclosure laws, and litigation that questions test scores as valid measures for selection–promotion. In short, tests and testing procedures have become controversial. The effects reach beyond the schools. Promotions and other test-based benefits to employees in industry and government are part of the debate.

This book has one principal focus—it asks whether and to what extent

The Rise and Fall of National Test Scores

Copyright © 1982 by Academic Press, Inc.
All rights of reproduction in any form reserved.
ISBN 0-12-068580-9

the controversy about school test results may be based on a tendency to simplify a very complex set of events. It addresses itself to the question, In what ways have school test scores been changing during recent years, and what factors probably account for those changes?

Before exploring this focus further and examining how this volume will approach the main question, consider the context in which the story unfolds.

Ever since the Soviet Union jolted America's identity as the world leader in science, engineering, technology, and the schooling that undergirds them, our educational system has become the object of steady scrutiny. Most Americans believe that schooling can and will make a difference for their children and, hence, their nation.

To judge the quality of schooling while a child is attending school poses a complex problem. The proper criteria will not be available until that remote time when school days are ended and the child acquires adult status. But parents, and the public in general, must make judgments here and now.

The only readily available evidence of how the schools are succeeding, other than teacher grades, is test scores. Teacher grades, by conventional agreement, are regarded as more a reflection of student characteristics (e.g., native ability and motivation to perform) than successful teaching. One could make the same assertion about test scores. Yet, in the area of standardized, normative testing, the disputants seem to have ignored this convention and have adopted the assumption that the schools are failing in their responsibility to teach effectively since certain test scores quite apparently are declining.

The nation expressed its desire to overtake and surpass Soviet technological training and education by a series of legislative acts. One, the Elementary and Secondary Education Act (ESEA) of 1965, combined the need to improve schooling with the desire to guarantee equal educational opportunities and futures for all children enrolled in America's schools. Hopes were high, and a feeling of general optimism prevailed. After all, we were making a huge investment in our space program, and early developments indicated that our scientific and technological abilities might make us the first nation to put men on the moon.

But there was some disquieting news. In 1966, the Educational Testing Service (ETS), which administers the Scholastic Aptitude Test (SAT) for the College Entrance Examination Board (CEEB), reported that for the first time in nearly two decades the national test score average showed a persistent decline that had begun 3 years earlier.

The famous Coleman Report (Coleman *et al.*, 1966) was issued at about the same time. It presented evidence that racial integration in America's schools was not leading to the expected improvement in minority students' performance. The following year, the *New York Times* announced under generous headlines that "City Pupils [were] Losing Ground in Reading and Arithmetic (Nov. 2, 1967)."

A few years later, another sociologist concluded that scholastic achievement has little to do with an individual's economic success. Christopher Jencks, in his widely read and quoted *Inequality* (1972), declared that economic success was poorly predicted by test scores and that such factors as on-the-job competence, personality, and just plain luck seemed to be the principal determiners of higher earnings. Were our hopes and expectations about American schooling misplaced?

These are a sampling of the views the public was exposed to over a period of about 15 years. Quite naturally, this nation responds actively to problems. The idea that our schools seemed to be failing did not go unnoticed. The reactions were quick in coming and broad in their applications.

The State of New York began an experiment with the so-called "truth in testing law," which was debated in Congress and may become a nationwide mandate unless some voices urging restraint are heeded (National Council on Measurement in Education, 1980). Such a law assumes that the test publishing industry itself, not merely those who use test scores, can harm the student candidates who take academic aptitude (selection) tests such as the SAT as well as those who take admissions tests for various types of graduate study. The implication is that the tests may be faulty in describing a given student's level of preparation, or, worse yet, that the company who administers the test may have committed errors in scoring or in designating the correct answers to the test questions and, thus, denied opportunities to trusting individuals.

Another implication is that the tests are biased against those groups who do poorly on them and that, somehow, student access to the test questions and keyed correct answers will lead to fairer examinations and admissions decisions. Interestingly enough, the first opportunity for students to request their tests, answers, and answer keys in New York State led to some surprising results. Only 7.6% of the candidates made the request. Of this group, most were in the top tenth of their high school class with grade point averages at or above 3.5 (on a scale of 0 to 4.0) and with SAT scores averaging 57% and 79% higher than average on the verbal and mathematical parts, respectively. Evidently, those groups for whom well-meaning advocates had urged the need for government intervention and protection seemed not to share the alarm nor to seek the protection (Council for Basic Education, 1980).

In addition to legislation designed to rectify perceived faults in aptitude testing, another set of mandates was quickly put into place. These were intended to deal more directly with the seeming problem of falling achievement test scores at the elementary and secondary school levels. Competency-based curricula and testing regulations were installed in many states. These moves, it was hoped, would hold the schools accountable for students' ability to perform at specified levels in various basic skills with particular stress on the mathematical and literacy areas. Early consequences of this movement suggest that knotty problems of determining satisfactory definitions of important, measurable activities have begun to plague curriculum planners. Only one of the problems is the sheer bulk of objectives. A completely satisfactory inventory of desirable competencies, unfortunately, would be so lengthy as to be impossible to attain in several lifetimes of schooling.

So, this period in American educational history has been marked by a quickening of activity to rectify perceived wrongs both in the schools and in the test publishing industry. Has the profession of education stood idly by and taken no part in the proceedings? Indeed not.

The decline of achievement test scores has been the topic of concern at a number of national conferences in recent years. In 1975, two such conferences were held—one to consider declines in school achievement, the other to consider declines in achievement test scores. The first, "The Decline in School Achievement," was held at Elkridge, Maryland, sponsored by the Edison Foundation and I/D/E/A. The second, held at the National Institute of Education in Washington, D.C., was titled "Declining Test Scores."

In 1976, the American Educational Research Association (AERA) and the National Council on Measurement in Education (NCME) co-sponsored a symposium at the AERA annual convention. Gilbert R. Austin, one of the editors of this volume, was co-sponsor with Lawrence M. Stolurow of the symposium. Much needed information was shared about the patterns of test score changes from several perspectives. (Several of the chapters in this book are by presenters at that national meeting.)

Recognizing the need for a concentrated, major effort to learn why college admissions test scores continued to decline year after year, the CEEB appointed an interdisciplinary study panel in 1976 to try to find some satisfactory answers. Willard Wirtz, former Secretary of Labor, chaired this group which issued its report the following year. Bruce K. Eckland, a member of that panel, presents a summation of this blue ribbon group's 1977 findings in Chapter 2.

Another outstanding activity of the educational research community was an invited address at the 1977 annual convention of AERA, delivered by Ralph W. Tyler, on the topic of declining test scores. Additionally, regional conferences and symposia, journal articles and reviews, and numerous books and pamphlets continue to reflect the national concern.

Many concerned educators, psychologists, sociologists, and other professionals have engaged actively in discussion, research, and debate. Some professionals assume either hostile or strongly positive stances in advancing their ideas; others are more moderate and scientifically dispassionate. The contributors and editors of this volume have tried to present material that will shed light on rather than enflame passionate issues. We believe the issues involved in the present controversy are of extreme importance to America's well-being. This country, indeed, the entire world, faces enormously compelling problems which need not be recited here. The nature of current global affairs demands that we use the powers of reason to their fullest. It follows, then, that the one institution that should nurture and provide for the development of rational thinking, the schools, needs constructive help in order to deal in a productive way with the current controversy that swirls about it.

Unfortunately, this controversy makes it extremely difficult to identify relevant facts and, moreover, offers little guidance in the search for pertinent information. This book presents some of the needed data and at the same time attempts to show how difficult it can be to get at them. It is hoped, too, that this presentation will act as a model for what is certain to be a continuing examination of a most critical problem.

We have taken the position that the best approach to this presentation is one that deals with the major issues and also recognizes the traditional organization of knowledge as presented in elementary and high schools. Hence, Chapters 4–7 address the questions surrounding trends in the basic subject matter areas of writing and literacy, reading, mathematics, and science test results, respectively. But also, since the changing test score phenomenon involves academic aptitude testing and college admissions, two chapters dealing specifically with those topics are included. One is data-based and factual (Chapter 2), and the other is concerned with the continuing issue of the special problems confronting racial minorities in American education (Chapter 9).

Since minimum competency achievement testing has become a reality in so many parts of the country, a chapter that deals specifically with this attempted solution also has been included (Chapter 10). Also, two research-oriented centers have been studying changes in pupil accomplishments over a fairly long period of time. The American Institutes for Research (AIR), under the leadership of John C. Flanagan, and the National Assessment of Educational Progress (NAEP), currently headed by Roy Forbes, have collected, systematically, achievement and other test data from young people across the country over a period of up to 20 years. These are among the best known data sources available to help answer questions about what American youth are achieving in academic as well as noncognitive areas, including the so-called basic skills. Chapter 3 was contributed by Flanagan, Chapter 4 by Forbes.

If patterns of test score trends differ in certain European cultures, a comparison of these differences to patterns in American schools should prove helpful to our understanding of the problem in this country. For this reason, the book includes the chapter by Thomas Kellaghan and George Madaus on test score trends in Great Britain and Ireland.

Popular opinion frequently points to eight societal forces as the most likely causes of test score declines. We asked our contributing authors to try, where appropriate, to relate their discussions to these factors. These factors can play significant roles in students' academic achievements and, therefore, in learning processes and test score outcomes. The factors are changes in:

1. Social policy concerning education (e.g., creating the Elementary and Secondary School Act, Head Start, and Follow Through)
2. The social milieu (e.g., television viewing habits, TV programs such as *Sesame Street* and *The Electric Company,* and the rise of mysticism and alternative cultures and religions)
3. The educational curriculum (e.g., the instruction of modern mathematics, new science programs, e.g., BSCS, PSSC, Harvard Project Physics)
4. School practices (e.g., open classroom, busing, token economies, contract systems, mainstreaming)
5. Impersonalization of society and schooling in terms of increased use and abuse of technologies
6. Democratization of schooling and the reduction of standards for both admissions and performance, thereby altering the population and sampling distributions of students in terms of preparation, interest, values, expectations, and standards
7. The social rewards provided by society to the more educated, better trained so that income and earnings become less related to educational experience and personal competence and more related to other social forces
8. Technical problems associated with the processes used in developing, revising, and using tests and evaluation instruments

The reader will note that the specific issue of whether the funding of education should be based on test results is absent from this list. The movement toward accountability measures, generally, although of substantial importance to educational practice and school organization, would be peripheral to the particular focus of this volume. Nevertheless, a concern for the effects of such sociopolitical actions will be felt by some of our contributors and may break into open expression occasionally. If it does, we will not feel compelled to suppress it. Indeed, it is of great importance in this presentation that the reader be reminded, from time to time, that the history of American education is a rich one and that what may seem unique today has been tried in the past and nearly forgotten. As an aside, it is interesting to note that the schools as such are not among the protagonists in the accountability subcontroversy. The advocates prefer to issue their arguments in the public forum rather than in the schools and universities. Nevertheless, some commentators are ever mindful of what school purposes ought to be, and they assert that where test scores show significant declines we would be well advised to examine the educational institutions for the source of the problems rather than merely censuring the instruments (Ebel, 1976).

We caution the reader to be aware of one ever-present limitation to an examination of this type, and several of the contributors allude to and, indeed, make explicit this particular constraint. The technique of testing, when focused on human behavior, presents a paradox. This puzzle emerges as one considers the fact that although mankind has used mental testing of some kind throughout recorded history, it rests finally upon rather shaky logical foundations. The testing procedure is of no particular long-lasting interest to those for whom the testing is conducted (teachers, administrators, etc.), or, at least, ought not to be. What *is* of central interest is the series of events that occurred before (i.e., teaching and learning) or that will occur after (i.e., the likelihood of success in further training or practice). But try as psychometricians may, they can never fully satisfy the wants of those they attempt to serve. In their own defense, test constructors design tests that rank students in an order similar to the way the students were ranked by a previously selected criterion. But these correlations are never perfect and, worse yet, selecting criteria is always a matter of human judgment. Hence, testing experts warn the users (administrators, admissions officers, etc.) to be cautious in making decisions. But are such warnings heeded? Sadly, the test is all too often seen to be at fault because some decisions made on the basis of individual scores are obviously wrong. And with a degree of justification, the user protests that the test is suspect since, in these cases, the results failed to satisfy their needs. But decisions must be made, and those who make them must often use imperfect test scores since they are usually the best data at hand.

In actuality, such protest is justified in a diabolical way, even though it seldom generates solutions. People need to understand the world around

them. The struggle to uncover the complex nature of human learning spans
the history of mankind. We hope that this book can contribute some useful
knowledge about the instruments that are used to assess and measure
human learning as it occurs in our schools. If it leads to some improvement
in school practice, we shall be most thankful.

REFERENCES

City pupils losing ground in reading and arithmetic. *New York Times,* November 2, 1967.
Coleman, J. S., Campbell, E. Q., Hobson, C. J., McPartland, J., Mood, A. M., Weinfeld,
    F. D., and York, R. L. *Equality of educational opportunity.* Washington, D.C.: U.S.
    Office of Education, 1966.
Council for Basic Education. A forecast of our own. *Basic Education.* September, 1980.
    Washington, D.C.: The Council for Basic Education.
Ebel, R. Declining scores: A conservative explanation. *Phi Delta Kappan.* December, 1976.
    Bloomington, Ind.: Phi Delta Kappa.
Jencks, C., Smith, M., Acland, H., Bane, M., Cohen, D., Gintis, H., Heyns, B., and
    Michelson, S. *Inequality: A reassessment of the effect of family and schooling in
    America.* New York: Basic Books, 1972.
National Council on Measurement in Education NCME Statement on Educational Admis-
    sions Testing. *NCME Measurement News.* 1980, *23(2),* 4–6.

Chapter 2

College Entrance Examination Trends

Bruce K. Eckland

Academic ability tests have been used in college admissions and placements during most of this century, though their roots may be traced much further back in our educational history. They have been developed and administered by various agencies for widely different purposes with the common goal of seeking to measure a high school student's basic aptitude and preparedness for college.

Modern admissions testing dates to 1900, when the College Entrance Examination Board (CEEB) was founded as a membership association by a small number of colleges, universities, and secondary schools who were concerned with the multiplicity of entrance examinations and the diversity of school curricula. During the first quarter of this century, the College Board administered, on a national basis, a series of standardized essay examinations covering different subjects. The tests were primarily used in the admissions process of private, eastern colleges, while public institutions in the Midwest and other regions tended to follow an admissions process governed by high school diploma or certificate (this was pioneered by the University of Michigan). In 1926, the College Board introduced a new multiple-choice Scholastic Aptitude Test (SAT) developed by Carl Brigham of Princeton University. The SAT was supplementary to the essay tests, and was used initially in the selection of scholarship candidates from schools not preparing students for the essay examinations.

Multiple-choice achievement tests developed by the College Board in

The Rise and Fall of National Test Scores

Copyright © 1982 by Academic Press, Inc.
All rights of reproduction in any form reserved.
ISBN 0–12–068580–9

cooperation with the American Council on Education (also connected with scholarship awards in 1937) came into regular use as part of the admissions process during World War II (1942) and subsequently came to be used not only in selection but also increasingly for placement purposes. To consolidate the nonprofit testing activities of the College Board, the American Council on Education, and the Carnegie Foundation for the Advancement of Teaching, the Educational Testing Service (ETS) was established in 1948 as a new and separate agency in Princeton, New Jersey, and it assumed responsibility for operational aspects of its founders' testing programs and for the conduct of research as well.

During the middle 1950s, the National Merit Scholarship Corporation was begun with the purpose of identifying high school students (usually in the eleventh grade) who would merit special commendation and economic incentive. It used a special scholarship qualifying test as its initial screening device. At about the same time, the College Board introduced the Preliminary Scholastic Aptitude Test (PSAT) as an instrument for use in the guidance and counseling of pre-college students. In 1971, the PSAT was combined with the National Merit test to become the PSAT/NMSQT, which now serves purposes of both early guidance and scholarship screening.

During 1959, still another national pre-college testing program was introduced with the founding of the American College Testing Program (ACT) in Iowa City. The ACT Assessment Battery serves essentially the same functions as the SAT and finds its heaviest use in the Midwest. A majority of eleventh graders who take the PSAT/NMSQT also now take either the SAT or ACT (or both) when they reach the twelfth grade.

Most students who go to college today take one or more of these tests. Most of the increase in the number of test takers occurred in the 1950s and 1960s. Those two decades were a period when colleges were expanding very rapidly due both to an increasing proportion of students in each age cohort planning to go to college and, as a result of the postwar baby boom, to the increasing size of each cohort in absolute number. Most of the open door colleges of the 1940s (e.g., the Big Ten universities) could not expand rapidly enough to admit everyone who applied. An acceptable means had to be developed for rejecting some applicants without directly penalizing those from underprivileged backgrounds (such as by raising tuition fees). The use of test scores, along with high school grades, was widely defended as being the only reasonable and fair procedure.

Evidence in defense of ability tests in college admissions comes largely from validity studies of such tests in predicting educational attainment, especially the academic performance of students in the classroom. Most past studies in this area focused on college freshman grades and found that the SAT and ACT are both valid predictors, although usually not quite as good as a student's high school grades or class rank. This finding was not unexpected considering the fact that high school grades and college grades are not only both related to aptitude but that a student's grade performance in both school and college also depends on motivation. Admissions tests, however, still have predictive power in college, independent of one's high school grades, partly because grades depend considerably on the academic and competitive environment of the particular high school a student attended. For example, a reasonably well-motivated and able student who graduates in the bottom half of a high school class composed predominantly of students above average in ability could be severely penalized if grade averages were the only criterion used in college admissions. Past studies on the predictive validity of test scores on college performance have been remarkably consistent ever since such studies were developed. According to a recent College Board report, this includes scores for minority students (Breland, 1979). And there is growing evidence that, perhaps due to grade inflation in our high schools, the predictive validity of both the SAT and ACT have been increasing in recent years. Since about 1974, the correlation between the ACT test and freshman grades at over 300 colleges has actually been as strong as the correlation between high school and college grades (Ferguson, in preparation).

It should be added, however, that ability tests are not only good predictors of who will do well or poorly in college, but they also are important predictors of who goes to college, who goes where, and who drops out or graduates. In the long run, in fact, it could be argued that these particular outcomes ultimately are more important to one's future than the grades one happens to receive in college. An exemplary study on this issue is the ongoing National Longitudinal Study (NLS) of the high school graduates of 1972, sponsored by HEW through the National Center for Education Statistics (NCES) and conducted by ETS, the Research Triangle Institute, and the U.S. Census Bureau.

Using an ability test not very dissimilar from the SAT or ACT, a recent summary of the first 30 studies to come out of the NLS project (Eckland & Alexander, 1980) reports that ability was not only a powerful predictor of who went to college immediately after high school but also of delayed entrants, as well as of who attended full or part-time. Ability also was one of the strongest predictors of who went to which college (as measured by the "selectivity" of the college), whether it was a two-year or four-year institution, and, in the case of blacks, whether the student attended a predominantly white or a traditionally black college. Among those who entered college, ability also was far more important than most other background factors in predicting who dropped out and who graduated on schedule. Among those who dropped out of college, ability was the strongest predictor of who eventually returned. This last but often neglected phenomenon is of considerable importance, given the fact that so many dropouts actually do return. Moreover, whether or not a student returns was found in the NLS not only to be strongly related to ability but almost wholly *un*related to the grades students obtained in college before dropping out. Finally, separate analysis by race found that the predictive power of the NLS test battery was generally just as strong for blacks as for whites at all points, from who goes to college to who goes where and who graduates. A recent ETS study funded by the College Board obtained similar but, in one respect, even more striking results (Wilson, 1978). The predictive power of admissions tests (in this case the SAT) on the long-term performance of college students (as measured by one's class standing and graduation four to seven years following matriculation) was found to be much stronger for blacks than whites.

In summary, there is no factor that has a stronger or more consistent effect in the college attainment process independent of race, social class, sex, or any other observed characteristic of high school students than individual ability as measured by standardized tests. For this and other reasons, any marked change in the distribution of test scores is likely to have a profound impact not only on who succeeds in college but on the life outcomes of young people. Although fluctuations in the SAT scores have been greater in some years than others, the trend since 1963 has been steadily downward. During the past 17 years, the average scores of all test takers have dropped by about one-half of one standard deviation, a statistically as well as substantively significant decline, and it appears that this decline still continues.

The SAT currently is given to well over one million high school students each year, a figure which represents over half of the total number going to college annually. It includes separate verbal and math scores, each reported on a scale of 200–800. Between 1952 and 1963 there were apparently random yearly fluctuations in the scores despite a very marked increase in the proportion of students taking the test (see Table 2.1). Between 1963 and 1980, however, the scores almost steadily declined, going from 478 to 423 on the verbal part and from 502 to 467 on the math part.

The 55-point drop in the score averages on the verbal part and the 35-point drop on the math part add up to a total drop of 90 points. As mentioned earlier, this represents about one-half of one standard deviation. But what does this really mean? As in most areas of the country, there is in my own state (North Carolina) a rather big difference between the selectivity of some institutions as compared to others, a matter that is

TABLE 2.1
SAT Scores and Population Changes Between 1951-52 and 1978-79

(a)	(b)		(c)	(d)	(e)
Testing Year	SAT Mean[a]		High School Graduates (in 1,000s)	SAT Test takers[a] (in 1,000s)	Percent of d/c
	Verbal	Math			
1951-52	476	494	1,196.5	81.2	6.8
1952-53	476	495	1,198.3	95.5	8.0
1953-54	472	490	1,276.1	118.1	9.2
1954-55	475	496	1,351.0	154.5	11.4
1955-56	479	501	1,414.8	208.6	14.7
1956-57	473	496	1,439.0	270.5	18.8
1957-58	472	496	1,505.9	376.8	25.0
1958-59	475	498	1,639.0	469.7	28.6
1959-60	477	498	1,864.0	564.2	30.3
1960-61	474	495	1,971.0	716.3	36.3
1961-62	473	498	1,925.0	802.5	41.7
1962-63	478	502	1,950.0	933.1	47.8
1963-64	475	498	2,290.0	1,163.9	50.8
1964-65	473	496	2,665.0	1,361.2	51.1
1965-66	471	496	2,632.0	1,381.4	52.5
1966-67	467	495	2,679.0	1,422.5	53.1
1967-68	466	494	2,702.0	1,543.8	57.1
1968-69	462	491	2,829.0	1,585.6	56.0
1969-70	460	488	2,896.0	1,605.9	55.4
1970-71	454	487	2,943.0	1,537.2	52.2
1971-72	450	482	3,006.0	1,459.9	48.5
1972-73	443	481	3,037.0	1,398.4	46.0
1973-74	440	478	3,069.0	1,354.0	44.1
1974-75	437	473	3,140.0	1,371.2	43.7
1975-76	429	470	3,150.0 (est.)	1,415.0	44.9
1976-77	429	471	3,147.0 (est.)	1,401.9	44.6
1977-78	429	469			
1978-79	426	466			
1979-80	423	467			

[a]Figures are for all SATs taken in a given testing year, including those at any grade level and those counted more than once if they repeated the test.

Source: College Entrance Examination Board, 1980, Personal Communication.

13

clearly recognized by the public and has a strong bearing on who goes where to college. For example, the average total SAT score of all entering freshmen at the four leading and predominantly white public colleges in North Carolina (NC State, Chapel Hill, Greensboro, and Charlotte) is about 985 today. In the six less prestigious but still predominantly white public colleges (Appalachian, East Carolina, Pembroke, Asheville, Wilmington, and Western Carolina), the average total SAT scores are about 848. This is a 137-point difference between these two groups of colleges. The national SAT score decline is almost (about two-thirds) as large as this. If readers have any first-hand knowledge of differences in the average level of competence of students attending the higher and lower status colleges in their own states, they may simply apply that difference to the SAT score decline and will not be much off the mark.

As early as 1975, the College Board and ETS received so many questions and so much public attention because of the unexplained decline in test scores that the two organizations appointed the Advisory Panel, a 21-member panel, to investigate and interpret the SAT score decline. Over the next two years, the Advisory Panel met in plenary sessions and work groups, and published its report, *On Further Examination,* in 1977. Most of what follows in this chapter is based on that report and on the 38 separate studies that the panel commissioned to be done. This chapter is also an update of the Advisory Panel's report, since some of the research studies recommended by the panel were not completed or initiated until after the report was published. Thousands of pages have been written on the subject by now, and I take full responsibility for what follows, especially my interpretation of some of the still unsettled issues. Most of the original reports that I will be referring to are readily available from the College Board.

Three basic questions were addressed by the Advisory Panel and will be addressed here. First, Has either the content of the tests or the type of training that students receive in school changed, thereby leading to the SAT score decline? Conceivably, the tests could simply have become more difficult over time or changes in educational philosophy and practice could be lessening their validity as relevant and practical measures of academic competence. Second, Have the types of students who take the SAT changed in any marked way? If the characteristics of those who go to college have been changing, there is a strong chance that score averages from one year to the next would either go up or down depending on who takes the tests. And third, If some or all of the above does not wholly account for the decline, then what does? Is it the quality of our schools, the family, or are there other major societal events at work?

THE CONTENT OF THE TESTS

Although the SAT was designed to be an unchanging measure of academic ability, it is quite possible that a current test score does not indicate exactly the same level of ability that a test administered ten or twenty years ago did. Since most of the questions are changed each year with each new edition of the test, it could have become harder (or easier) over time, despite ETS's best equating and scaling procedures.

At the Advisory Panel's request, two technical studies were conducted in order to check ETS's methods. In one of these, 3174 students from 66 high schools were given *both* the 1963 and 1973 editions of the SAT (Modu & Stern, 1977). The results confirmed the results of earlier analyses indicating that the test has not become more difficult but has become easier. The decline since 1963 in SAT scores has actually been about 8–12 points *larger* than reported earlier (about 100 instead of 90 points).

The panel also attempted, but could not answer, a much more difficult question about changes in the "relevancy" of the test to the school learning process. At issue here were not changes in the formal curriculum or changing standards (that came later in its report) but whether or not different learning and communication processes had come into use in the schools that the SATs no longer reflected. For example, it is possible that conventional standardized tests like the SAT do not measure competencies that are developed in the schools through increasing reliance on tape recorders, film strips, or other "kin-audiovisual" forms of instruction. Lacking any evidence to answer questions of this kind, the panel concluded that while such issues deserve further inquiry, the current tests should not be compromised: "If the value base of the SAT is accepted as being solely the prediction of college academic performance, the critical fact is that the test's predictive validity is actually somewhat higher than it used to be" (CEEB, 1977, p. 10). Throughout the panel's deliberations, it was pointed out that college admissions tests like the SAT were never designed to measure the effectiveness of the nation's schools but to help predict how well students will perform academically in college.

WHO TAKES THE TESTS

As mentioned earlier, the 1950s and 1960s saw an extraordinary rise in the number of young people going to college, along with increasing proportions taking the SAT and ACT tests. Much of this increase actually came about long before the average scores started to decline. As Table 2.1 shows,

81,000 SAT tests were taken in 1952, with the number rising to 933,000 by 1963. However, during this period there was no steady rise or fall in the *average* scores. Was nothing really happening to affect test scores in the 1950s or were there offsetting factors?

For example, it is entirely possible that dramatic compositional changes occurred during this period with increasing proportions of low ability students taking the test each year. No decline in the average SAT test results would have shown up, however, if there was an equally dramatic upward spurt in the total academic competence of each new cohort. Or, just the reverse could have happened. That is, students graduating from our high schools in the 1950s could have been becoming steadily less competent, on the average, as measured by any standardized ability tests (if they had been given). However, it did not show up on the SAT because during those years it was mainly the more selective colleges that required the tests and mostly just the higher ability students who took them. But this is all conjecture and, due to an almost total lack of evidence, was not a matter taken up by the Advisory Panel.

Nevertheless, the panel clearly recognized the possibility that an increasing percentage of young people staying in school and taking the tests could have brought about a drop in the *average* scores, and quite probably did, in the 1960s when the decline first appeared. Some data already were available from a series of Freshman National Norm studies made by ACT (Ferguson & Maxey, 1976) and from the Student Profile studies (Maxey, 1976) made by ACT, both of which tended to support the hypothesis. Yet, the best evidence came from an analysis the panel commissioned that compared biographical data and the test scores of two high school senior classes. The two studies used in the analysis were the national Project TALENT study, conducted in 1960, and the NLS of 1972 discussed earlier (Beaton, Hilton, & Schrader, 1977). The Beaton *et al.* (1977) study made it possible to equate the reading exams given to the 1960 and 1972 high school senior classes, to identify who took the SATs, and thereby to estimate exactly how much of the drop in the scores of SAT test takers during this period could be explained by a shift in high or low ability students in each class who went to college and who took the SAT.

The results clearly showed that proportionately more students with lower scores began going to college and were taking the exam during the 1960s. Proportionately more blacks with low scores accounted for a few points in the total SAT decline, more women taking the tests accounted for a small part of the drop in the math scores, and more students from lower socioeconomic backgrounds accounted for a much larger share of the overall drop. In addition to a clear expansion of educational opportunity, part of the score decline also was directly attributed to the marked increase in the percentage of test takers going to less selective colleges with open admissions policies, especially the number applying to two-year institutions. In summary, the panel found that between two-thirds and three-fourths of the score decline between 1963 and the early 1970s could clearly be accounted for by compositional changes in the SAT-taking group, part of which was associated with changes in who was planning to go to college and, independent of a student's college plans, part of which was associated with changes in who took the SAT.

Although the panel itself did not investigate the issue, a study by Flanagan (1976) showed that little if any of the drop in test scores could be accounted for by the large proportion of each age cohort during the 1960s who completed high school. Between 1964 and 1970, for example, the proportion of young people staying in school to the twelfth grade increased from two-thirds to three-fourths. However, Flanagan found during this period, as to be discussed in the next chapter, no uniform decline in the test performance of high school students.

It also should be pointed out that not all of the decline by 1972 was due simply to more low ability students taking the SAT. As the panel noted, some part of the early decline was due to a small drop in the proportion of high ability students electing to go to college (Peng, 1976), a phenomenon that could not be explained.

But what happened in the 1970s? Was the continued decline over the past decade also largely due to changes in the test-taking population? More precise data on the background characteristics of the test-taking population became available in the 1970s from the Student Description Questionnaire, administered by ETS, which was first distributed with the SAT in 1971–1972 and is now completed by a majority of all SAT test takers. These data confirmed the continuing existence of compositional change in the test-taking group during the 1970s, much like the changes that had taken place in the earlier years, but not nearly as great as before. The evidence consistently pointed to an emergence of what the panel called "pervasive" forces that were causing the decline and were unrelated to who happened to take the test.

Score averages continued to go down in the 1970s in all types of schools (private, public, large, and small) among all students including those from high and low socioeconomic backgrounds, both men and women, and whites as well as blacks. The panel also found that during the 1970s the actual number of high scorers on the SAT (above 600 on either the verbal or math sections) had dropped greatly, from 189,300 in 1970 to 108,200 in 1976 (Jackson, 1976). Were many of the high ability students not taking the SAT any longer or were these students simply not doing as well as earlier cohorts on the tests? In the panel's opinion, it was mainly the latter.

The apparent emergence of a very real decline in the academic ability of young people, including that of even the most able students, was confirmed in a special study that the panel made on the SAT scores of some 1500 valedictorians and salutatorians in 145 high schools (Donlon & Echternacht, 1977). The research showed that while no significant drop occurred in the average test scores of such students during the 1960s, after 1970 their scores declined by about the same amount as those for the total population, a phenomenon that could have little to do with any continuing change in the composition of the test-taking population in the 1970s. Other factors must have been at work.

Fully aware that declining SAT scores alone probably do not warrant generalizations about the abilities of all young people, the panel also reviewed statistical evidence from other widely used tests, including the ACT, the PSAT/NMSQT, and ETS's Achievement Tests and postgraduate school admissions tests. In summary, the ACT showed a decline comparable to the SAT since the mid-1960s, with most of the drop having taken place in the 1970s. National samples of all eleventh graders showed substantial stability in score averages of the PSAT/NMSQT exam in the 1960s and early 1970s, but revealed drops in both the verbal and math sections since 1973 that were almost exactly parallel to the SAT decline. (This finding in particular supports the panel's conclusion that the continuing SAT score decline in the 1970s was not mainly due to compositional changes.) Some of the College Board's Achievement Tests, covering 14 different subject areas and taken in conjunction with the SAT, have also shown score declines, while others have shown increases. The results on the Achievement Tests are exceedingly difficult to interpret, however, because fewer students in the 1970s had taken them and because there were marked changes in their use during the score decline period. Lastly, graduate and professional school admissions exams have had mixed results. Although the Law School Admissions Test and the Medical College Admissions Test showed increases in average scores, both the Graduate Record Examination (GRE) and the Graduate Management Admissions Test showed declines similar to the SAT.

Based on all of the evidence at hand, the Advisory Panel thus concluded that the score decline since 1963 developed in two distinct stages. The panel's own studies and a comparative analysis of the SAT with various other tests all confirm the fact that most of the score decline during the 1960s was due to changes in the composition of students going to college and students taking the SAT and ACT tests. During the 1970s, however, most of the decline was caused by more "pervasive" forces and probably only about one-fourth of the total subsequent drop in scores is traceable to compositional changes in the test-taking population. What these forces might be, exactly, is the next subject.

As the panel concluded its work, the SAT score decline looked as though it were ending. Between 1976 and 1977 there was no drop in the verbal part and only a one-point drop in the math part. Three years later, however, the verbal section dropped another six points and the math section dropped two points. In my opinion and probably that of most others who served on the panel, the continuing drop in test scores reflects patterns of social change in our society at large.

SOME POSSIBLE PERVASIVE FORCES

Having concluded that most of the SAT score decline during the 1970s and at least part of it during the 1960s was not due to changes in the composition of students taking the test, the Advisory Panel reviewed about 50 different theories that were brought to its attention. They ranged from the rise of school crime and violence to the negative consequences of the post–World War II baby boom on the American family. In discussing most of these theories the panel freely expressed its opinion on their soundness, but also admitted that the search for causes was "essentially an exercise in conjecture" (CEEB, p. 25). It seldom was possible to assign to any given theory or factor the degree of influence it may have had on the score decline or to even be certain that it had had any effect on SAT scores at all.

As research scientists generally are well aware, a correspondence or correlation between any two variables (x and y) does not necessarily mean that x is causing y. The correlation may simply be coincidence (due to chance or random error), or spurious (where both x and y are dependent on some antecedent event), or causally connected but in the reverse direction (y causing x). In discussing the various hypothesized theories of the score decline and weighing the evidence, the panel tried to consider all of these possibilities. Given the heterogeneity of the panel's membership and a general lack of good evidence, its opinions often were mixed. Nevertheless, the panel was not reluctant to state its convictions and consistently did so in its final report.

Before considering the panel's conclusions, one other point should be mentioned that was not clearly stated in the panel's report but was clearly in the minds of its members. That is, any particular factor causing the SAT score decline, and we assume there was more than one factor, could have been operating as early as the 1940s on the lives of those who took the tests in the 1960s, or as early as the 1950s on those who took the tests in the

1970s. That is, we have no way of knowing exactly when the decline in the measured ability of any given cohort of high school juniors or seniors may actually have begun. Although most speculation tends to focus on the high school years or the period of adolescence, the crucial events could have occurred in infancy or in the early school grades. The panel had no way of knowing, and thus considered all possible factors. As listed below, they fall into three general areas:

In the School
   The academic curriculum
   Other learning standards
   The teachers and teaching methods
The Broader Learning Context
   Parents
   Television
   Other societal events
Student Motivation

The Academic Curriculum

Much of the attention of observers outside the panel has focused on the decline of "basic" required courses in our high schools, particularly in the English and verbal skills area, and the increase in "electives." Although the evidence suggests that these events indeed did occur, it is unclear just how much or exactly in what manner they contributed to the SAT score decline.

In the widely circulated Harnischfeger and Wiley report (1975), it was found that in the period between 1971 and 1973, high school enrollments in English courses nationally had dropped about 11% and courses in advanced English about 50%. However, ETS data from the Student Descriptive Questionnaire, completed by the SAT takers, showed no such drop. From year to year, about 90% of these students consistently reported having taken four English courses, although no one knows exactly what kinds of courses and this probably is where the problem lies.

Evidence from some states, such as Massachusetts and California, points to the belief that nonacademic, or elective, courses are partially responsible for test score declines. For example, in Massachusetts it has been reported that between 1971 and 1976 those high schools showing substantial enrollment increases in the two most common English "speciality" courses, that is, Science Fiction and Radio/Television/Film, also showed the largest SAT score decline. And in California, it was reported that English composition class enrollment fell by 77% between 1972 and 1975, while English electives in Children's Theater, Mystery and Detective Story, and Executive English nearly doubled.

Based on evidence of this kind, the panel concluded that even though the exact nature of the shift in content from traditional English courses to the newer electives is unclear, the shift has probably contributed to the decline in SAT verbal scores to some extent. The panel concluded that the important factors in this respect presumably are: "(1) that less thoughtful and critical reading is now being demanded and done, and (2) that careful writing has apparently about gone out of style [CEEB, 1977, p. 27]."

There also has been a related development in the sharp decline of high school student enrollment in foreign language courses. This decline was strongly emphasized to the panel by specialists in this area since there is an unquestionable correlation between the number of foreign language courses that students take and their SAT verbal scores. However, as the panel pointed out, the causal nature of this relationship is unclear. In all probability, such a relationship exists mainly because students with higher verbal scores tend to take more foreign language courses. Three observable facts supported this conclusion: (a) Students who take foreign language courses do just as well on the SAT math section as they do on the verbal section, which disproves the assumption that foreign language courses mainly contribute to the student's knowledge and understanding of what is being measured in the verbal section; (b) although the proportion of all students taking foreign languages in high school clearly has declined, there has been no such reduction since 1973 among students who take the SAT; and (c) the SAT score decline since 1973 has been at least as large among students with foreign language exposure as among those with no exposure.

Under the topic of high school curriculum, the panel also gave consideration to the possibility that the decline on the math section of the SAT may be due to some kind of change in the number and type of high school mathematics courses being taken today. However, no firm answer was reached on the matter. There has not been the same kind of proliferation of electives in mathematics courses that there has been in the verbal skills area. Student enrollments in high school mathematics, moreover, have not gone down nearly as much as enrollments in English courses, and although the scores on the verbal and math sections of the SAT dropped by about the same amount in the 1960s, the decline on the math section was significantly less than the decline on the verbal section in the 1970s. These three points, when taken together, would seem to reinforce the possibility that the drop in the SAT verbal scores was indeed at least partly due to changes in the academic curriculum of our high schools.

Two other possible effects of school curriculum on score declines were given considerable attention by the panel, but its conclusions were negative.

One of these involved a possible upward shift in the proportion of high school students focusing on more vocational or technically oriented majors than liberal arts. Such students, it was thought, may be going to college and lowering the SAT score average. Although those in the vocational curriculum do score much lower on the SAT than those in the academic curriculum, the percentage enrolled in vocational courses who take the SAT is very small. This percentage has actually not changed in recent years, and the score decline has been almost identical in both groups. The panel reviewed the related possibility that the increasing proportion of high school students partaking in experimental programs or out-of-school training may also have contributed to the score decline, but it appears that this is not the case. Test takers who reported working outside of school 1–15 hours per week not only had higher SAT scores than those not working at all, but their scores have been declining less than the overall average.

Other Learning Standards

The academic curriculum is only one measure of learning standards that the Advisory Panel considered. Others include the rise in absenteeism, grade inflation, social promotion, reduced homework, and a marked drop in the level of difficulty of textbooks. The panel concluded that in all five of these areas there has been a lowering of educational standards and that some of them are probably factors in the SAT score decline, although the extent of their relevance is unknown.

Absenteeism (daily attendance) seriously worsened, especially in the high schools, during the 1960s and 1970s. Absentee rates of about 20 or 25% in some schools are no longer unusual. The panel placed as much of the blame for this on the home and the community as on the schools, and it assumed that attendance drops have had a corrosive effect on students as a whole and, therefore, on test scores.

Grade inflation essentially means that a much higher percentage of students today are receiving "A" and "B" grades than ever before, which is clearly confirmed from various reports. The panel, however, doubted that this actually contributed to the SAT score decline since "the distribution of SAT takers among the various *percentiles* on high school grade records apparently has not changed substantially [CEEB, 1977, p. 29]." In other words, there is still a substantial correlation between high school performance as reflected in grades and SAT scores. In this instance, I think the panel's interpretation of the data could have been wrong. As mentioned earlier, there is now evidence from ACT that the predictive power of admissions tests on college performance rose significantly between 1965 and 1979, while the predictive power of high school grades dropped (Ferguson, in preparation). In response to grade inflation, student motivation could certainly be on the decline, at least in so far as course grades are concerned.

Automatic promotion from one grade level to another was, thought the Advisory Panel, a more significant factor than grade inflation. The main problem is not the consequences for those who are promoted when their work does not warrant it, for most of these students will not go to college anyway. Rather, it is commonly assumed that the better students suffer when teachers have to deal with classes containing too many who are not up to the same level of instruction.

The amount of homework assigned to students at all grade levels is thought to have been on the decline for more than a decade, although the panel could find no reliable evidence of just how much is given now as compared to the past. One of the problems is that so many students now claim to do most of their homework while watching television. In any case, the panel concluded that reduced homework assignments are related to the SAT score decline.

The last item in this section, a drop in textbook standards, was given special attention by the panel. It commissioned an investigation of the readability and level of difficulty of the reading and history textbooks that were most often used in the first, sixth, and eleventh grades over the past 30 years (Chall, 1977). The study found that most of the current eleventh-grade texts are at a ninth to tenth-grade reading level and that more textbook space is now taken up by pictures, larger print, and shorter words, sentences, and paragraphs. The panel concluded that the tendency of textbook publishers to "make it simple" is clearly a lowering of educational standards and another possible cause of the SAT score decline.

Any attempt to adapt mass education to the needs and desires of *all* students, as reflected in lower grading standards, more social promotion, a reduction in homework, and easier textbooks, has the authentic purpose of meeting the needs of some students while unintentionally shortchanging others—a problem that I will come back to later.

School Teachers and Teaching Methods

Given the fact that some members of the Advisory Panel were themselves school teachers or administrators, the panel admitted that it was a poor critic or judge of the extent to which, if at all, the score decline may reflect a decline in teacher competence. It noted that the rise in student enrollments during the 1960s and early 1970s caused a sharp increase in the demand for teachers, which resulted in a significant drop in teacher experience—from an average years-of-experience for elementary teachers of about 13 in 1961 to 8 by 1971. On the other hand, the panel noted that the educational levels of teachers actually rose during this period. Moreover, pupil–teacher ratios in both elementary and secondary schools apparently decreased.

The Advisory Panel did, however, discuss in its meetings (but purposely did not do so in its final report) the possibility that part of the score decline could be due to a reduction in the academic competence of teachers. It seems there are two reasons why the issue was avoided. One, which is mentioned in the panel's report, is that many state legislatures today have enacted (or are in the process of doing so) bills creating statutorily prescribed ability tests for the hiring of new teachers, an issue in which neither the College Board nor ETS probably wants to be involved. The other reason, discussed at the meetings, was that no reliable norms on the National Teachers Examination (NTE) over time are available due to the fact the states that use the test have changed from one year to the next.

It seems to me that the quality of teachers could very well have declined over the past two decades, and this could have contributed to the SAT score decline. Moreover, the situation could be getting worse. Since it was not part of the panel's report, I will only spend a short time here on the issue.

First, over the past two decades there are reasons to believe that the academic competence of teachers has been on the decline. In theory, this would be true because increasing numbers of women, who for generations have occupied a majority of the nation's teaching jobs, have been gaining access to higher status and more lucrative occupations. As a result, less able women are now teaching. Historically, teaching has been a woman's profession, but one that, I believe, largely due to the women's liberation movement, has become less attractive to many women. There is nothing new about the fact that schools offering teacher education programs typically graduate the least academically talented college students. However, there is strong evidence, based on past and recent SAT, ACT, GRE, and NTE tests taken by new teachers (Weaver, 1979), that the situation has become steadily worse.

Second, there is strong evidence that the competence of teachers affects the competence of students. Many educators want to forget the Coleman Report (1966) because it assigned most of the explanation for the variance in students' ability to the family rather than to the schools. What many also forgot was its central finding about what had the greatest effect on student performance among all of the numerous school characteristics that James Coleman investigated; namely, a 30-item vocabulary test administered to the teachers. In other words, the most important measurable impact of the schools on the measured competence of their students was not the curriculum or a school's resources, but the quality of its teachers.

And third, I would argue that the situation can be expected to get worse and not better over the next decade, especially among blacks. Affirmative action in higher education is leading an increasing proportion of the members of all minority groups, both women and racial minorities, into more lucrative occupations. Teaching was once the only higher-than-average status job in which competent females and blacks could be guaranteed employment. Now that the situation has markedly changed, and is being reinforced by the continuing unionization of teachers (an issue that the panel also purposively ignored), there is little hope for the future unless the conditions governing who is motivated and hired to teach are fundamentally changed.

Aside from the quality and training of teachers, the panel made a particular effort to explore changes in teaching methods, such as the movement in many schools toward more "open classrooms." These investigations and the review of the literature in this area, however, proved inconclusive. All of the major longitudinal studies (of elementary school children) have failed to find any substantial or consistent relationship between school achievement, as measured by standardized tests, and the level or type of innovation employed in the classroom.

The Role of the Family

As mentioned earlier, most educators do not acknowledge the proposition, which is now supported by numerous studies, that family background explains substantially more of the variation among students in academic ability than do differences in the quality of our schools or teachers. Although the Advisory Panel's report also did not acknowledge this proposition, it gave a considerable amount of attention to changes in the structure of the American family and to the possibility of a causal relationship between these changes and the SAT score decline.

Of the various views on changes in the family over the past two decades, the first to be considered was Zajonc's (1976) popular theory that score declines can be wholly explained by the rise in family size and the decreasing proportion of first-born children in the population. There is no question that, due to the post-World War II baby boom, the decline in test scores after 1963 is almost perfectly concomitant with the drop in the number of first-borns after 1947. In other words, throughout the score decline period until 1979, the proportion of test-taking high school seniors from large families steadily increased, while those who were first-borns steadily decreased. Assuming, as Zajonc does, a strong causal relationship between either family size and ability or birth order and ability, the SAT score decline could conceivably be explained fully by the baby boom.

However, Breland's (1977) paper written for the panel finds that the relationship between birth-order and academic ability was simply not an important factor in the SAT score decline, presumably because the later, less able siblings chose not to take the SAT. Therefore, while the theory may be valid, it could not explain more than 6 scale points of the SAT score decline.

If the rise in family size did indeed contribute to part of the score decline, what remains unknown is why the correlation exists at all. Although the panel did not go into this part of the issue in its report, my own reading of the research literature suggests a number of alternative explanations. One is the "confluence" theory of Zajonc and others that various factors in the home environment contribute to the intellectual development of children and that these factors become especially diluted when children are spaced close together in age. Another is the possibility that if intelligence is largely inherited and if the intelligence of parents is negatively correlated with their fertility, then the baby boom could have produced part of the observed correlation between family size and the score decline. Or, as some studies have found, most of the relationship between family size and the measured ability of children disappears when social class is controlled, which means that the correlation could simply be spurious.

Other changes in the family also were considered by the panel, including the fact that the number of children with divorced parents had doubled over the past decade, the fact that the number living with only one parent or none had been increasing at the rate of over 300,000 each year, and the fact that 40% of all women with children under age 6 were now working—this figure had been rising for decades and still has not stopped. Although the panel admitted that any causal relationship between these changes and the score decline is unknown, most members believed that it was probably more than coincidence.

The panel also noted parallel weaknesses in the traditional relationships between parents and the schools. Changes in educational practice, as well as in the family, have contributed to a strain on the teacher–parent relationship. Yet, the panel repeatedly stressed the point that, although matters of home learning and parent–teacher cooperation are no doubt of critical importance and probably have had a bearing on the score decline, the gaps in our present state of knowledge are simply too great to know what the key factors really are and how much weight to give them.

The World of Television

Once again, it cannot be proved that television contributed to the decline in SAT scores, and, if it is a factor, no one knows how much. Nevertheless, the panel was convinced that television watching is one cause of the SAT score decline. "By age 16 most children have spent between 10,000 and 15,000 hours watching television, more time than they have spent in school. When they reach the first grade, their average watching time is between 20 and 35 hours a week; this usually peaks at about age 12 [CEEB, 1977, p. 35]." The rise of television accompanied the baby boom after World War II, and by 1965, when the SAT scores were beginning to drop, 95% of all American homes had television sets.

Although several studies, reviewed by the panel, on the correlation between ability test scores and television viewing have been conducted that generally find the hypothesized negative relationship, none are longitudinal or sophisticated enough to be conclusive. Two reasons, however, are given as to why television may have been a cause of the score decline. One is that a considerable amount of the time children now spend watching television used to go into homework (and into reading and writing). The other is some evidence of the apparent fact that reading a textbook or an examination involves a "linear, verbal, logical" function which operates in the left hemisphere of the brain. In contrast, watching television mainly involves a "simultaneous, visual, affective" function which occurs in the right hemisphere of the brain. If this is true, it is at least possible that a lot of television watching could affect the neural mechanisms of the mind and, over time, could lower an individual's ability to handle verbal materials. The panel viewed the prospect as alarming but as another area where we stand only on the frontier of knowledge.

Other Societal Events

In addition to the school, the family, and television, numerous other events in the society at large were considered, some of which most panel members were inclined to believe related to the score decline but, again, could not be proven. In particular, it was noted that in the late 1960s and early 1970s we, and especially young people, went through a period of national disillusionment due to the country's involvement in a divisive war, political assassinations, and burning cities. Although it is unknown exactly what effect these events had on the motivation and test scores of students, it may be more than coincidence that it was during the early 1970s that both the SAT and ACT scores declined most sharply.

The panel also considered the possibility that desegregation and "forced busing" may have contributed to the score decline. While there is documentary evidence in some cities that busing has had a negative impact on the educational process, such disruptions generally appear to be of short duration. The panel also pointed out that school desegregation has actually been restricted to a fairly small proportion of the nation's student population, which means that it could not have had much effect on the total SAT score averages. It also should be noted, even though the panel did not do so, that the one region of the country that has actually been forced to desegregate most of its schools, the South, has had a less than average drop on the SAT, 30 points between 1972 and 1979 as compared to 40 points for the nation as a whole.

Student Motivation

Of the 50 or more plausible explanations that the panel considered, loss of "motivation" was the most elusive. The topic naturally came up in many of the different contexts already discussed. For example, is reduced "achievement motivation" a possible cause or consequence of absenteeism, of social promotion in our schools, or easier textbooks? Is it a cause or consequence of television? Or is it one more element in the assumed national disillusionment of youth that began a decade or more ago?

A summary of the research literature on motivational factors in the score decline, commissioned by the panel (Winter, 1977), showed, again, the general complexity and lack of relevant information that we have on this subject. We do not even know whether "n Achievement" (the need to achieve) levels in the youth population have risen or declined over the past decade. Many people just assume that they declined and, according to the panel, "It seems plausible speculation that as opportunities for getting into college have widened there may have been less concentration of student efforts on preparing for college entrance examinations [CEEB, 1977, p. 38]." In other words, the incentive and motivation to achieve and do well on the tests could have lessened due to a drop in competitiveness in college admissions.

But does the panel's simple and seemingly plausible statement actually coincide with the facts? I don't believe so. It is true that due to the transformation of teachers' colleges into liberal arts colleges and to the rise of junior and community colleges, access to college has increased for everyone. However, this does not appear to have actually slackened the need for most students to do well on college admissions tests. On the contrary, just the opposite appears to have occurred. As stated in the introduction, undergraduate enrollments have markedly risen over the past several decades and this has required the more prestigious colleges and universities in most states to become more selective and to place more (not less) emphasis on test scores. Moreover, also as noted earlier, the evidence on college entrance examinations indicates that in recent years they have become more (not less) predictive of college performance and therefore are being given more (not less) weight in the admissions formula. In my own discussion with hundreds of students over the past few years, it is assumed by a majority of them (wrongly) that the tests were actually more important than their high school grades for gaining admission to a selective college. The motivation to do well on the tests, therefore, has probably been increasing, not decreasing. Of course, "*n* Achievement" specialists might then argue that any increasing emphasis on the tests would increase a student's level of anxiety and thereby tend to "freeze" or inhibit test performance. While it is generally agreed that reducing anxiety would improve the performance of some small percentage of the test-taking population, it is not known by how much or to what extent, if any, the level of anxiety has risen.

The earlier statement, quoted from the panel's findings, that students do not concentrate as much as they once did on preparing for college entrance examinations implies that not only are they less motivated but they spend less time in actual preparation. Again, I would argue that the evidence we now have points in the opposite direction. Over the past decade there has been a marked increase in the development of special high school programs that prepare students for a college admissions test. According to a recent College Board report (Alderman & Powers, 1979), about one-third of the schools in the Northeastern states now offer programs specifically designed to improve a student's performance on the SAT verbal section. The report also found, like most past studies, that special training of this kind can indeed boost students' verbal performance by about 8 points. Thus, it appears that students actually are spending more (not less) time in preparing for the SAT and, to some extent, benefit from the extra effort, yet the scores have continued to decline.

Given all of these problems and unknowns, the panel, as one might assume, concluded that it was not competent to judge the merits of the motivation argument, but believed it was an area that warranted much more study. The panel also considered many other kinds of factors that could have caused or resulted from a possible decline in student motivation, such as the participation (or lack of participation) of students in various extracurricular high school activities, the rise in school violence, juvenile delinquency, alcoholism, and drug addiction. While common sense tells us that at least some of these behavior patterns probably are causally related to the score decline, the panel again had no answers.

THE PANEL'S SUMMARY AND RECOMMENDATIONS

Of the three broad areas covered by the Advisory Panel (outlined earlier), the first was strongly conclusive, the second was moderately conclusive, while the third was filled with nuances, qualifications, and doubts.

First, in so far as the technical aspects of the SAT score decline are concerned, the evidence strongly supported the panel's conclusion that the decline is not due to the tests having become harder. In fact, they have become easier, and the actual decline is 8–12 points more than what has generally been reported to the public. The panel also concluded that despite the decline in score averages, the predictive validity of the tests, as measured by college grade performance, has not decreased but, rather, has slightly increased in recent years.

Second, a substantial part of the SAT score decline, about half of it overall, could be attributed to compositional changes in the group of students taking the test rather than to any change in the abilities of this age cohort or to high school graduates as a whole. Part of this change involved the increased proportion of each age cohort who went to college, and part of it involved the increased proportion of those planning to go to college who took the SAT. In other words, coincident with the extension and expansion of educational opportunity in the United States, increasing numbers of lower ability students began taking the tests. The panel concluded, however, that most of this compositional change occurred in the 1960s. By the 1970s, the SAT-taking population had begun to stabilize with respect to economic, ethnic, and social background. The continuing drop over the past decade, therefore, appears to have been caused by more "pervasive" social forces.

Third, exactly what these forces are is largely unknown. Is the score decline partly a result of the lowering of educational standards, or is it a result of social forces outside the schools and over which educators have no control, or both? The panel concluded that in regard to most of the dozens of hypotheses it considered, there was no conclusive answer. Nevertheless, the evidence in support of some theories tended to outweigh that of others. The following summarizes the panel's major conclusions:

1. In regard to the high school curriculum, the panel agreed that the adding of more "electives" and the apparent drop in enrollment in basic English courses have contributed to the large score decline on the verbal section of the SAT. Critical reading and careful writing have essentially gone out of style. The panel could not find any support, however, for the hypothesis that the SAT score decline is causally related to the fact that the number of students taking foreign languages also has defined or to the fact that more students today are taking experimental and out-of-school jobs where they are working part-time.

2. There is clear evidence of diminishing standards in education, any one of which could be a cause of the score decline, and they come in inter-

related forms: the tolerance of increased absenteeism, grade inflation, automatic promotion, a reduction of homework, and lower reading levels of textbooks.

3. While teaching methods, such as the trend of more "open classrooms," have markedly changed, there is no evidence that innovations in this area have had either a positive or negative effect on standardized test scores. The panel also tended to rule out the possibility of a correlated decline in teacher competence. On this last point, however, I disagreed with the panel, having found evidence both that the test scores of teachers have declined and that the ability of teachers has a significant effect on the cognitive development of students.

4. Changes in the role and social structure of the American family have accompanied the score decline. One of the most often cited theories in this area is the assumed relationship between the decline in test scores and the increase in family size following the post-World War II baby boom. Although one of the panel's commissioned studies found evidence in support of this proposition, it could not explain more than 6 points of the SAT score decline. Other observable changes in the family also were noted, such as the increasing proportion of married women working outside the home and the rise in divorce rates. However, the effect of these changes on the score decline could not be determined.

5. Because of the advent of television, which now consumes more of the average child's time by age 16 than school does, most children's learning now develops through viewing and listening than through the traditional modes of reading and writing. The panel strongly supported the proposition that television detracts from homework, that it competes with school, and that it has contributed to the score decline.

6. The largest drop in SAT scores occurred in the early 1970s, which was a period of national distraction due to a divisive war, political corruption, and city riots. The panel strongly suspected that there was a real relationship between the disruption throughout the nation and the score decline during this period. The 1970s was also a period of school desegregation and forced busing, which the panel, in all probability, correctly assumed had little impact on the score decline.

7. Lastly, the panel considered the possibility of a drop in student motivation and its relationship to the score decline. In its report, the panel assumed that this was a plausible explanation on the supposition that students now concentrate less on the tests since the opportunities for getting into college have widened. I have argued, however, that just the reverse has probably occurred. The more prestigious col-

leges in most states have placed more (not less) emphasis on admissions tests; the tests have become more (not less) predictive of college performance; students typically believe that the tests are more (not less) important than their high school grades in being admitted to the college of their first choice; and increasing numbers of motivated students are voluntarily enrolling in the special training programs that are now being offered in the high schools and that are specifically designed to improve a student's test performance.

In summary, the Advisory Panel concluded that the SAT score decline indeed was real and was actually larger than reported to be (i.e., after taking into account that the tests have become somewhat easier). A small part of the decline in the 1970s, but most of it in the 1960s, was due not to a lowering of the measured ability of high school seniors, but to the increasing proportion of lower ability students going on to college and taking admissions tests. The remaining part of the decline, especially in the 1970s, was probably the consequence of numerous pervasive forces operating in the schools and in our society. What these forces actually are is largely, although not entirely, a matter of conjecture.

Since most of the score decline (56 of the 90 points that dropped since 1963) occurred in the 1970s, and since only about one-fourth of the drop during this period can be attributed to compositional changes in the groups who took the test, there does indeed appear to have been a serious deterioration of the learning process in America, and there is no evidence, at least among the current crop of high school students, that it has stopped.

Given the fact that almost every member of the panel was in one way or another associated with the educational establishment, most of its attention was focused on the schools. However, it also became readily apparent that far more information on the schools in relation to the score decline was available than on other social institutions (e.g., the family). I believe that one of the most important recommendations of the panel, therefore, was the need for much more information than is presently available on the circumstances surrounding test-takers outside of their school settings. In the panel's opinion, the correction of whatever is currently wrong in the cognitive development of children will only come from the collaboration of teachers, students, parents, and the broader community.

The panel also warned against perceiving the SAT or any other standardized tests as the sole thermometer of the health of our schools, students, or families, because test scores tell us nothing about honesty, and integrity, or other things that matter more.

The panel was equally adamant about the need for maintaining diversity in the educational process and for recognizing the diverse needs of students. The American educational system serves an extremely broad constituency. The panel refuted suggestions such as that all students be uniformly held in a grade until they reach a common standard, that they be suspended from school as a rigid penalty for absenteeism, that they be overloaded with homework, or that "electives" be removed from the curriculum. What we do need is a system that can be more varied without being watered down; that is, one that promotes majoring in special fields without consequent effects on verbal and quantitative skills. "The lowering of teaching sights is the wrong answer to whatever may have been the consequences of the expansion and extension of educational opportunity [CEEB, 1977, p. 47]."

Several studies recommended by the panel have already been completed by the College Board, and some were used in this chapter (Alderman & Powers, 1979; Breland, 1979; Wilson, 1978). While committed to studies on the SAT itself, the College Board is in full support of the panel's recommendations for further study on the passage point between secondary and higher education and on "circumstantial evidence" outside the schools that may be related to the score decline.

In this regard, there is one study now in progress that deserves mention. This is "High School and Beyond," a longitudinal study of the sophomore and senior classes of 1980 and a direct parallel to the continuing National Longitudinal Study of the class of 1972 (discussed earlier). Both national surveys are sponsored by NCES. James Coleman is directing "High School and Beyond," and one of his first tasks, using tests identical to those administered in 1972, will be to compare the 1972 and 1980 test results and to explain the continuing but inexplicable score decline.

ACKNOWLEDGMENTS

This report is based mainly on the work of the CEEB–ETS Advisory Panel on the SAT Score Decline. I especially wish to thank George H. Hanford, current president of The College Board, for his helpful comments on an earlier draft. The members of the panel were Willard Wirtz (chairman), Harold Howe, II (vice chairman), Bernard C. Watson, Ralph W. Tyler, Ledyard R. Tucker, Vivian H. T. Tom, Robert L. Thorndike, Barbara Thompson, Thomas W. F. Stroud, Rosedith Sitgreaves, Wilbur Schramm, Katherine P. Layton, Owen B. Kiernan, H. Thomas James, Matina S. Horner, Edythe J. Gaines, Frank W. Erwin, Bruce K. Eckland, Luis C. Cortes, Sandra A. Clark, and Benjamin S. Bloom.

REFERENCES

Alderman, D.L., and Powers, D.E. *The effects of special preparation on SAT-Verbal scores.* New York: College Entrance Examination Board, 1979.
Beaton, A.E., Hilton, T.L., and Schrader, W.B. *Changes in the verbal abilities of high school*

seniors, college entrants, and SAT candidates between 1960 and 1972. New York: College Entrance Examination Board, 1977.

Breland, H.M. *Family configuration effects and the decline in college admissions test scores.* New York: College Entrance Examination Board, 1977.

Breland, H.M. *Population validity and college entrance measures.* New York: College Entrance Examination Board, 1979.

Chall, J.S. *An analysis of textbooks in relation to declining SAT scores.* New York: College Entrance Examination Board, 1977.

Coleman, J.S., Campbell, E.Q., Hobson, C.J., McPartland, J., Modd, A.M., Weinfeld, F.D., and York, R.L. *Equality of educational opportunity.* Washington, D.C.: U.S. Office of Education, 1966.

College Entrance Examination Board. *On further examination: Report of the advisory panel on the scholastic aptitude test score decline.* Princeton: College Entrance Examination Board, 1977.

Donlon, T.F. and Echternacht, G.J. *A feasibility study of the SAT performance of high-ability students from 1960 to 1974 (Valedictorian Study).* New York: College Entrance Examination Board, 1977.

Eckland, B.K. and Alexander, K.L. *The national longitudinal study of the high school senior class of 1972.* In Alan C. Kerckhoff, *Longitudinal Perspectives on Educational Attainment.* Greenwich, Connecticut: JAI Press, 1980.

Ferguson, R.L. *The predictive validity of the ACT assessment* (ACT Research Report No. 81). Report in preparation.

Ferguson, R.L. and Maxey, J.E. *Trends in the academic performance of high school and college students.* Iowa City: American College Testing Program, January 1976.

Flanagan, J.C. *Changes in school levels of achievement: Project TALENT ten- and fifteen-year retests.* Paper presented at the annual meeting of the American Educational Research Association, San Francisco, April 22, 1976.

Harnischfeger, A. and Wiley, D. *Achievement test score decline: Do we need to worry?* Chicago: CEMREL, Inc., 1975.

Jackson, R. *An examination of declining numbers of high-scoring SAT candidates.* New York: College Entrance Examination Board, 1976.

Maxey, E.J. *Trends in the academic abilities, background characteristics, and educational and vocational plans of collegebound students.* Iowa City: American College Testing Program, May 1976,

Modu, C.C. and Stern, J. *The stability of the SAT-Verbal score scale.* New York: College Entrance Examination Board, 1977.

Peng, S.S. *Some trends in the entry to higher education.* Paper presented at the annual meeting of the American Psychological Association, Washington, D.C., 1976.

Weaver, W.T. In search of quality: The need for talent in teaching. *Phi Delta Kappan*, September, 1979, pp. 29–46.

Wilson, K.M. *Predicting the long-term performance in college of minority and non-minority students.* New York: College Entrance Examination Board, 1978.

Winter, D.G. *Motivational factors in the SAT score decline.* New York: College Entrance Examination Board, 1977.

Zajonc, R.B. Family configuration and intelligence. *Science,* April 16, 1976, 227–236.

Chapter 3

Analyzing Changes in School Levels of Achievement for Men and Women Using Project TALENT Ten- and Fifteen-Year Retests

John C. Flanagan

The findings being reported here are based on Project TALENT and related studies developing from it. Project TALENT was conceived more than 20 years ago as a national survey of students in grades 9, 10, 11, and 12. Plans were made for follow-ups for each class at intervals of 1, 5, 10, and 20 years after graduation. The sampling design was developed to include a sample composed of about 5% of the secondary school students, with each sampling unit weighted so that the total sample would be representative of all the secondary students. About 400,000 students in these schools spent two school days taking tests and responding to questionnaires and inventories.

The three follow-ups already completed have provided a great deal of information about the quality and characteristics of the nation's educational institutions. They have also furnished the basis for an educational and career guidance program that is entitled *Planning Career Goals* (AIR, 1975–1977). This survey and the several follow-ups have made possible a very large number of special studies related in only an incidental way to the major purposes originally planned for project TALENT. The data base has been used by well over a hundred persons representing a variety of disciplines to investigate specific problems.

Before proceeding with a discussion of specific results, certain essentials for the valid analysis of changes in school levels of achievement need to be emphasized. The first necessity is to be completely sure that any differences

35

Copyright © 1982 by Academic Press, Inc.
All rights of reproduction in any form reserved.
ISBN 0-12-068580-9

noted are not due to the sampling procedures used in the two periods being compared. Precise and comparable samples are an essential prerequisite to any valid comparisons of trends. The second procedural factor is the emerging importance of sex as a confounding variable in studying trends in school achievement and the need for studying the performance of each sex separately. The question of sex as a confounding variable is to be briefly discussed now. The problem of sample comparability will be postponed until later in this chapter when the results of 1970 and 1975 retests are described.

To illustrate the large sex differences in certain test results in the 1960 National Project TALENT Survey, a few examples are cited. An inspection of the results, from information tests in more than 30 specific fields, indicates that girls in the secondary schools in 1960 were definitely better informed on such matters as music, home economics, health, clerical work, theater and ballet, and foods. They also were slightly better informed than the boys about art, the Bible, colors, and etiquette. On the other hand, the boys were definitely better informed on matters concerning social studies, mathematics, the physical sciences, the biological sciences, aeronautics and space, electricity and electronics, mechanics, farming, sports, law, engineering, foreign travel, the military, hunting, fishing, outdoor activities, and sedentary games. Boys and girls were about equally well informed with respect to literature, architecture, journalism, photography, "practical" knowledge, and accounting, business, and sales.

Looking at the interests expressed by these twelfth-grade boys and girls, we find a number of differences. Occupations that five times as many boys as girls claimed they would "like very much" include civil engineer, toolmaker, auto mechanic, electrician, fireman, aeronautical engineer, electronics technician, bricklayer, riveter, electrical engineer, mining engineer, repairman, railroad brakeman, truck driver, mechanical engineer, machinist, welder, and carpenter. Five times as many girls as boys said they would "like very much" to be a : tailor or dressmaker, dietitian, airline hostess or steward, social worker, office clerk, elementary school teacher, nurse, secretary, librarian, typist, beautician, and switchboard operator. Thus, for this list of 122 possible occupations, we find overwhelming preferences for 30 occupations. There are 18 greatly preferred by male and 12 greatly preferred by female twelfth-grade students.

The 205 occupational and activity items included in the interest inventory were grouped into 17 scales. On these scales, differences (of about half a standard deviation or more) between the mean scores of twelfth-grade boys and girls, favoring the boys, were found for nine scales: physical sciences, engineering, mathematics, and architecture; public service; sports; hunting and fishing; business/management; mechanical/technical; skilled trades;

farming; and labor. The girls' scores were about one-half standard deviation or more higher than the boys' for five scales: literary/linguistic; social services; artistic; musical; and office work. Only small differences were observed for the remaining three scales: biological sciences; sales; and computation.

Although the self-descriptive personality scales were not found to have substantial predictive value for later roles and activities, they did show some differences in male and female patterns of response. Girls described themselves as having significantly more social sensitivity ("I never hurt another person's feelings if I can avoid it"), more tidiness ("I am neat"), and more culture ("I tend to have good taste"). Differences in their self-descriptions of sociability, impulsiveness, vigor, calmness, leadership, self-confidence, and mature personality tended to be small.

Another way of quantifying, at least to some degree, the cultural differences between twelfth-grade high school boys and girls in 1960 is in terms of their reports on participation in various types of activities. The activities that at least 15% more boys than girls reported they participated in "often" included: hunting or fishing; hockey, lacrosse, handball, boxing, wrestling, track, and field events; mechanical or auto repair; playing baseball, football, or basketball; retail store work, delivery; building models (airplanes, ships, etc.); cabinet making and woodworking; making or repairing electrical or electronic equipment; and metal-working. The activities that twelfth-grade girls reported a great deal more activity in than did the boys were: cooking, sewing, knitting, crocheting, and embroidering.

Similarly, there were substantial differences between the boys and girls in terms of the courses taken between grades 9 and 12. The boys reported taking more courses in science, mathematics, and shop or vocational subjects, while the girls reported taking more courses in commercial subjects. For example, in this 1960 sample, 37% of the boys as compared with 20% of the girls took at least four semesters of science; 45% of the boys as compared to 24% of the girls took at least four semesters of mathematics; 17% of the boys and 4% of the girls took shop or similar vocational courses; but only 8% of the boys as compared with 33% of the girls took at least four semesters of commercial courses. There were only small differences in the proportions of the two groups taking English, social studies, and foreign languages.

What effect did these primarily culturally inspired sex differences in preferences and activities have on the development of the cognitive abilities of the young people in this study? The survey results for grades 9, 10, 11, and 12 are shown in Table 3.1 for the 10 cognitive ability measures found to discriminate best between those entering various career fields. At the

TABLE 3.1
Mean Scores on Ten Ability Tests for Project TALENT Students
in Grades 9, 10, 11, and 12

	Grade 9	Grade 10	Grade 11	Grade 12	Raw Score Differences Between Grades 9 & 12	Standard Deviation Grade 9	Differences in terms of Standard Scores Between Grades 9 & 12
Vocabulary							
Males	16.0	17.6	19.2	20.4	4.4	5.92	.74
Females	15.2	16.6	18.0	19.3	4.1	5.60	.73
English							
Males	72.4	76.2	80.0	82.6	10.2	14.77	.69
Females	79.2	82.7	86.0	88.7	9.5	13.11	.72
Reading Comprehension							
Males	24.6	27.5	30.8	33.0	8.4	10.98	.77
Females	26.1	29.0	31.5	33.6	7.5	10.35	.72
Creativity							
Males	7.7	8.4	9.4	10.3	2.6	3.79	.69
Females	7.2	7.9	8.6	9.2	2.0	3.54	.56
Mechanical Reasoning							
Males	11.2	12.1	12.9	13.5	2.3	4.04	.57
Females	7.8	8.4	8.6	8.8	1.0	3.37	.30
Visualization in Three Dimensions							
Males	8.1	8.7	9.3	9.7	1.6	3.29	.49
Females	7.3	7.8	8.2	8.5	1.2	2.90	.41
Abstract Reasoning							
Males	8.0	8.6	9.2	9.6	1.6	3.16	.51
Females	8.0	8.5	8.9	9.2	1.2	3.12	.38
Quantitative Reasoning							
Males	7.4	8.1	9.0	9.8	2.4	3.45	.70
Females	7.0	7.6	8.2	8.8	1.8	3.28	.55
Mathematics							
Males	9.2	10.0	11.6	12.4	3.2	4.18	.77
Females	9.4	9.7	10.1	10.2	.8	3.85	.21
Computation							
Males	19.4	23.2	27.9	31.6	12.2	27.92	.44
Females	26.9	28.5	31.1	33.7	6.8	21.24	.32

Source: Flanagan, J.C. et al, 1964.

ninth-grade level, using standard score units to eliminate the effect of the raw score metric, the boys were found to score much higher than the girls on the Mechanical Reasoning test and definitely higher on a nonverbal test, Visualization in Three Dimensions, which has been found to be a good predictor of mechanical ability and other skills requiring visualization of objects. Though the differences are smaller, the boys had significantly higher scores on the Vocabulary, Creativity, and Quantitative Reasoning tests. The ninth-grade level girls showed their greatest superiority on the English exam, which tested spelling, capitalization, punctuation, usage, and effective expression. A somewhat smaller but substantial difference favoring the ninth-grade level girls was observed on the Computation test. The girls also show significantly higher scores on the Reading Comprehension test. On the Introductory Mathematics test, the girls' scores were very slightly higher than the boys', and on the Abstract Reasoning test there was no difference.

It is interesting to observe how these comparisons changed at the upper levels. In the 1960 sample, the English and Computation test scores showed the same substantial difference, favoring the girls for grades 10, 11, and 12. On the Vocabulary test, the boys' slightly higher achievement was maintained at all grade levels. On the Reading Comprehension test, the small difference favoring the girls was only half as large at the upper grades as at the lower levels. The small difference favoring the boys on the Creativity test became twice as large in the twelfth grade. The very large difference favoring the boys on the Mechanical Reasoning test was maintained, with the gains between ninth and twelfth grades being twice as great for boys as for girls. The fairly large difference on the Visualization in Three Dimensions test, shown by the boys in the ninth grade, was also found to be greater in the twelfth grade. For the Abstract Reasoning test, on which the performance of boys and girls was the same in the ninth grade, the boys showed a mean score in the twelfth grade that had a difference one-third larger than that of the girls when compared with the ninth grade score. The small difference favoring the boys at the ninth-grade level became a fairly large difference by the twelfth grade on the Quantitative Reasoning test. The most dramatic change in differences was that in the Introductory Mathematics test scores. At the ninth-grade level, the mean score for girls was very slightly larger, but at the twelfth-grade level, the boys' mean score was half a standard deviation higher than the girls'. The Computation test also showed a similar though relatively smaller increase for the boys. The substantial difference favoring the ninth-grade girls almost disappeared by the twelfth grade.

These mean-score comparisons of grade level groups in 1960 are of considerable interest in reflecting male–female differences and changes in these differences at various grade levels. The comparisons between grade levels are confounded by differential dropout rates and therefore do not provide valid measures of growth. To correct for this difficulty, about 10,000 students who participated in Project TALENT in the ninth grade were retested with some of the tests 3 years later when they were in the twelfth grade (Shaycoft, 1967). The results are shown in Table 3.2.

As in the norms comparison in Table 3.1, the differences between grades 9 and 12 in terms of standard scores were greater for the boys than for the girls on all tests except the English test. Also, the magnitude and direction of the differences between the males and females in the ninth grade were very similar to those for the larger group. Practically the only difference was that in the smaller sample the ninth-grade boys had a slightly higher mean score on the Abstract Reasoning test than the ninth-grade girls (rather than the same mean score). In terms of the differences between ninth- and twelfth-grade mean scores, the retests of the same students showed significantly smaller differences on the English test and the Reading Comprehension test. This tends to confirm other evidence that high school dropouts are mainly deficient in verbal skills. There is a corresponding but somewhat smaller relative gain shown on the retests for the Mathematics and Quantitative Reasoning tests, indicating that dropouts also accounted for some of the differences between the scores for the ninth- and twelfth-grade students on these tests.

The greatest gains by the boys between grades 9 and 12 appear to be due primarily to the types of courses taken by the boys as compared to the girls. As noted earlier, many more boys than girls took courses in mathematics, science, and social studies. The fact that the girls took more English courses is reflected in their greater gains on the test in this field. The gains shown by boys significantly exceeded those of the girls in Mathematics, Visualization in Three Dimensions, Quantitative Reasoning, and Mechanical Reasoning. Another point to be noted regarding this retest study was the fact that 4 of the 10 tests indicated that students recalled some of the special types of items used. Tests that showed this type of "practice effect" included Creativity, Visualization in Three Dimensions, Abstract Reasoning, and Mechanical Reasoning. Such effects appeared to be minor for the other 6 tests. An article reviewing some of these findings in *Women's Education* (Flanagan, December 1966) said, "These findings suggest that the education of girls is not adequate to prepare them for future activities and roles . . . . Probably the best thing a parent can do is help the child learn more about the world of work, including the requirements for and financial advantages of some of the kinds of work for which she could prepare herself. Also more girls, perhaps, might be urged to take work that requires logical reasoning and problem solving."

TABLE 3.2
Comparison of Mean Scores on Ten Ability Tests for Males and Females
Tested in Both the 9th and 12th Grades

	Grade 9	Grade 12	Raw Score Gain Grades 9-12	Standard Score Gain Grades 9-12
Vocabulary				
Males	17.2	21.3	4.0	.73
Females	16.4	20.2	3.8	.66
English				
Males	75.8	83.4	7.6	.51
Females	81.5	89.0	7.5	.55
Reading Comprehension				
Males	26.6	32.9	6.3	.60
Females	28.1	34.1	6.0	.59
Creativity				
Males	8.3	11.6	3.3	.87
Females	7.8	10.6	2.8	.79
Mechanical Reasoning				
Males	11.9	14.3	2.4	.61
Females	8.2	9.8	1.6	.47
Visualization in Three Dimensions				
Males	8.6	10.4	1.8	.58
Females	7.9	9.1	1.2	.42
Abstract Reasoning				
Males	8.8	10.2	1.4	.47
Females	8.5	9.7	1.2	.42
Quantitative Reasoning				
Males	8.0	10.1	2.2	.64
Females	7.6	9.2	1.5	.48
Mathematics				
Males	10.1	12.8	2.7	.66
Females	10.2	11.0	.8	.20
Computation				
Males	25.1	34.2	9.0	.45
Females	30.4	36.5	6.0	.37

Source: Shaycoft, M.F., 1967.

The data described above were collected in 1960 and 1963. In 1970, just 10 years after the initial Project TALENT survey, the administrators in 20% of the schools included in the original survey were asked to have their eleventh-grade students give us 1 hour of their time. The schools invited to participate were selected using the original 10 strata identified at the time the Project TALENT sample of schools was selected. A total of 12,722 eleventh-grade students in 134 of the original national sample of schools participated in the study. The results for the schools in each of the 19 strata were weighted to make the representation from that stratum the same as for the national sample in 1960. Each student took the Project TALENT Reading Comprehension test and answered 36 items that were selected from the 1960 form of the Project TALENT Student Information Blank.

The reading comprehension scale was identical to the scale contained in the 1960 TALENT battery. It consisted of eight passages to be read, with each passage followed by several five-option, multiple-choice questions testing comprehension of the passage rather than mere ability to recognize words. Passages dealt with social studies, natural sciences, and literary content. Poetry as well as prose passages were included. When answering the questions, students were permitted to refer back to the passages. Extremely difficult vocabulary was avoided. In general, the words were within the first 15,000 entries on the Thorndike–Lorge List (1944). It was felt that reading comprehension was a particularly good ability to tap because of the effectiveness of reading as a predictor of general school success and because of the attention that has been paid to improving reading skills in educational programs of the past decades.

Reading comprehension is independent of specific course content. It is usually not formally taught beyond the primary grades but is assumed to develop as a result of an individual's total educational experience.

The comparisons indicated that the boys in the eleventh grade in 1970 scored about .5 point higher than those tested in 1960 and the girls scored about .3 higher. In both years the scores of the girls were about a point higher than those of the boys.

It should be noted that the proportion of the relevant age groups attending the eleventh grade increased from about 77% in 1960 to about 85% in 1970. Since previous studies of the Project TALENT group indicated that dropouts had substantially lower scores in Reading Comprehension, the gains for students with comparable ability can be assumed to be significantly larger over this period. The difference between the proportion of males and females dropping out in 1960 as opposed to 1970 was small, and there was also a slight tendency for the male dropouts to have lower Reading Comprehension scores than the female dropouts (a number of whom dropped out because of marriage or pregnancy).

A few other reports from this study are relevant to the present discussion. For one, the students in 1970 reported that they spent less time studying than the students in 1960 reported. In 1970, 66% of the boys and 53% of the girls indicated that "on the average I study less than ten hours a week including study periods in school as well as studying at home." These figures were 6 percentage points larger than reported by the boys in 1960 and 4 percentage points larger for the girls. Their reports on reading books were very similar for the two periods with about 28% of both boys and girls indicating that they had read eleven or more books in the preceding twelve months. There was only a slight change in their reports on school attendance over this period. In 1960, 12% said they had been absent as many as 15 days in the last school year as compared with 10% for the 1970 group. Eleventh-grade boys and girls were both selecting occupations more appropriate to their abilities in 1970. Although in both years 50% of the boys reported that they planned to graduate from college, the girls having such plans increased from 35% in 1960 to 43% in 1970. There was also a comparable shift by the girls out of commercial courses of study and into college preparatory courses.

The five-year follow-up of the 1960 Project TALENT sample indicated that only 12% of the eleventh-grade males and 22% of the eleventh-grade females planned on the same career five years after finishing high school that they were planning when in the eleventh grade. It was also found that in 1960, many of the eleventh graders selected occupations very inappropriate to their ability levels. To check on the apparent appropriateness of the occupational choices of eleventh graders in 1970, an analysis was made of their abilities as shown by the Reading Comprehension scores of the groups planning occupational careers in each of 35 fields.

In general, the Reading Comprehension scores of the girls planning careers in various fields suggest that the girls in 1970 were a little more realistic than the boys. However, among the girls who showed major shifts toward some of the socially oriented fields requiring graduate training, these new plans did not seem to be so well grounded in a realistic appraisal of their own abilities as did those for the boys. Many of the girls planning careers in dentistry, engineering, law, mathematics, physical science, medicine, political science or economics, social work, sociology or psychology, and nursing appeared to have unrealistic goals in terms of the Reading Comprehension scores known to be required to enter training for these fields.

It is very encouraging to note that the eleventh-grade male students who indicated careers as biological scientists, lawyers, mathematicians, physicists, political scientists or economists, sociologists, and pharmacists scored, in 1970, much more like students who might be expected to enter these occupations than was true of the 1960 group. In most of the groups mentioned, the scores are at least 5 points or half a standard deviation higher than the average scores in 1960. Similarly, the scores for the boys planning to enter such fields as skilled workers, structural workers, and barbers or beauticians tended to be lower in 1970 than they were in 1960. This, again, is in accordance with realistic expectations for these students. Only the groups of boys planning careers as dentists, engineers, and social workers had Reading Comprehension scores which suggested that many of their plans were unrealistic.

In the spring of 1975, a second opportunity was provided to test students in some of the same grades in a number of the schools that participated in the 1960 national survey. As part of the standardization of a new battery of tests based on the Project TALENT findings, about 1800 students in grades 9, 10, and 11 were given the 10 Project TALENT tests.

Comparing student performance in the same schools over a 15-year period eliminates sampling problems in terms of school selection. However, trends in the socioeconomic level of the students served by these schools needed to be controlled. The school principals were asked to report observed changes in the communities using a 5-point scale. Differences of more than 10 percentile points were given values of either 1 or 5 on the scale and differences of 5 to 10 percentile points were assigned the intermediate values of 2 and 4. Differences smaller than that were assigned the middle value. These estimates were used to make small adjustments in the means for those schools where a change was reported. Four of the 17 principals estimated that the quality of the community was slightly lower, 10 reported it to be about the same, two reported their communities were slightly higher, and one principal reported that the socioeconomic status of the community served by his school was higher in 1975 than in 1960.

To get an estimate of the stability of the overall means for boys and girls on the various tests obtained in the 1975 sample, the 17 schools in the sample were randomly divided into two halves of about equal size (8 schools and 9 schools). Appropriate small corrections were made in each school mean for each of the 10 tests in accordance with the principal's estimates of changes in the character of the community served by the school between 1960 and 1975. These changes varied between a few hundredths and one- or two-tenths of a point. Mean scores on the 10 tests were then obtained for each sex group in the three grades (9, 10, and 11) for the two comparable samples. To obtain two independent estimates of each of the 60 mean score values, the mean scores from the 1960 and 1975 students of one sex at a specific school were weighted in accordance with the number of students from that school in this subgroup in the 1975 survey. To make these differences between 1960 and 1975 mean scores comparable for all measures, each difference, in terms of raw scores, was divided by the standard deviation for the 1960 scores. This procedure produced 60 pairs of independent estimates. The product moment correlation coefficient for these two estimates from random halves of the total sample was obtained. This was found to be 0.81. This coefficient provides an estimate of the consistency of the results to be expected from these half-size samples. Correcting, by using the Spearman–Brown formula to represent the entire sample being used, yields a coefficient of .90.

Thus, the results from this sample as shown in Table 3.3 can be regarded as quite dependable in spite of the relatively modest sizes of these samples. To make the raw score differences between 1960 and 1975 as shown in Table 3.3 more meaningful, they have been shown as percentile differences in terms of tenth-grade norms. As a first step in examining Table 3.3, it seems desirable to compare the 1960 results for this sample with those in Table 3.1. Nearly half of the 1975 sample were in the tenth grade with about equal numbers from grades 9 and 11. Therefore, the 1960 results for these three grades combined (as shown in Table 3.3) were compared with the tenth-grade norms for all schools in 1960 (as shown in Table 3.1).

It is clear that the 1960 results for these schools were quite close to those for the complete national sample in 1960. The mean scores of the students from these 17 schools were only a bit higher for these 10 tests than for all schools. These differences are about one-tenth of a standard deviation for all tests except the Creativity test. They are between one- and two-tenths of a standard deviation higher than the national sample on the Creativity test. Also, in terms of male–female differences in these 17 schools' mean scores on these 10 tests in 1960, the results are very closely similar to those obtained by the tenth-grade students in the total sample of schools.

Turning to changes between 1960 and 1975, there are several interesting general trends and some fairly large changes in the differences between male and female cognitive ability measures. Looking at the general trends first, the scores for 1975 students are significantly lower than for the 1960 students for the Vocabulary, English, Quantitative Reasoning, and Computation tests. The scores on the Reading Comprehension test are very slightly lower in 1975, and the scores on the Introductory Mathematics test are very slightly higher. Two tests show significant gains over this 15-year period—Creativity and Abstract Reasoning. The girls have improved more on the Creativity test and the boys have improved slightly more on the Abstract Reasoning test. The remaining two tests, Mechanical Reasoning and Visualization in Three Dimensions, show slightly lower scores for the boys and significantly higher scores for the girls.

This latter finding suggests that the movement to expand the interests and career opportunities for women is having significant impact on today's high school girls. Although the males still show much more ability in deal-

TABLE 3.3

Comparisons of 1960 and 1975 Results in Grades 9, 10, and 11 of
Students in 17 Secondary Schools Using the Same Project TALENT Tests
(For males, N-1975 = 871; for females, N-1975 = 926)[a]

| | Raw Score 1960 | Raw Score 1975 | Changes Between 1960 and 1975 | | |
			Raw Score	Standard Score	10th Grade Percentile Difference
Vocabulary					
Males	18.5	15.7	-2.8	-.48	-17%
Females	17.3	15.5	-1.8	-.32	-11%
English					
Males	77.3	73.4	-3.9	-.26	-12%
Females	84.5	79.7	-4.8	-.34	-16%
Reading Comprehension					
Males	28.8	28.4	- .4	-.04	- 1%
Females	29.8	29.1	- .7	-.07	- 2%
Creativity					
Males	9.1	10.0	.9	.22	8%
Females	8.4	10.1	1.7	.46	16%
Mechanical Reasoning					
Males	12.4	12.2	- .2	-.05	- 2%
Females	8.5	9.2	.7	.20	7%
Visualization in Three Dimensions					
Males	8.9	8.8	- .1	-.03	- 1%
Females	7.8	8.1	.3	.10	4%
Abstract Reasoning					
Males	8.7	9.5	.8	.26	11%
Females	8.7	9.4	.7	.22	8%
Quantitative Reasoning					
Males	8.5	7.8	- .7	-.20	- 8%
Females	8.0	7.2	- .8	-.23	- 8%
Mathematics					
Males	10.5	10.7	.2	.04	2%
Females	9.9	10.3	.4	.10	3%
Computation					
Males	25.7	18.7	-7.0	-.27	-17%
Females	30.8	26.9	-3.9	-.19	-11%

Source: Flanagan, J.C., 1976.
[a]Copyright 1976, American Educational Research Association, Washington, D.C.

ing with mechanical and spatial materials, the significant reduction in this difference during the past 15 years strongly suggests that this (apparently culturally induced) difference can be expected to change even more in the near future.

In attempting to find a reasonable explanation for some of the other changes in average scores of boys and girls over this 15-year period, it seems desirable to examine the probable effects of two factors. The first is the selective factors changing the composition of these groups in the tenth grade in 1960 and in 1975. The number of dropouts declined during this interval. In general, those dropping out of school in 1960 were from the lower ability levels. This trend should produce a decrease in mean scores over the period. The comparison of norm and retest results from the 1960 and 1963 surveys suggests that this factor will have its greatest effect on scores on the English, Reading Comprehension, Mathematics, Quantitative Reasoning, and Vocabulary tests. Three of these tests do show significant decreases, but two—Mathematics and Reading Comprehension—do not.

The second possible factor influencing these changes in test performance relates to the topics emphasized in the curriculum. Over the period from 1960 to 1975, there was a shift in emphasis in many schools away from memorizing the rules of grammar and from drilling on the simple operations of addition, subtraction, multiplication, and division, but in favor of a greater amount of time spent on developing basic skills such as reading comprehension and basic thought processes such as those exemplified in the Creativity and Abstract Reasoning tests. The increase in the number of students in the college preparatory curriculum over this period may be responsible for the greater acquisition of mathematics by the 1975 groups. This shift in courses probably also contributed to reading comprehension scores by emphasizing the use of academic materials.

Clear evidence of the trend toward increased participation in mathematics courses is provided by a recently released report by Jane M. Armstrong, one of the members of the Education Commission of the States. Her report, *Achievement and Participation of Women in Mathematics: An Overview* (1980), reports that 59% of twelfth-grade boys and 55% of twelfth-grade girls in the fall of 1978 indicated they had taken at least two years of mathematics. It will be recalled that only 45% of the twelfth-grade boys and 24% of the twelfth-grade girls reported having taken at least two years of mathematics at the time of the Project TALENT retest in 1963.

The recent study also provides some new comparative data on the performance of males and females on various tests of mathematical and spatial abilities. A sample of 1788 students in the twelfth grade in the fall of 1978 was subdivided into groups of boys and girls who had taken various amounts of mathematics in high school. To supplement this group, the results obtained from testing about 24,000 17-year-old students in the spring of 1978 were subdivided according to the amount of mathematics studied in high school. When the extent of formal mathematics study was controlled in this way, only small differences, mostly favoring the males, were found between the mean scores for boys and girls on tests of computational skills and algebra. However, for both studies, the boys did better than the girls on the problem-solving tests, with most of the differences being statistically significant.

The survey in the fall of 1978 also included a test of spatial ability. Unfortunately, this was a version of the Minnesota Paper Form Board test, involving only two-dimensional spatial visualization, and therefore the results cannot be compared to those obtained using Visualization in Three Dimensions (mentioned earlier). On the two-dimensional spatial visualization test, the 13-year-old girls scored 5 percentage points higher than the boys (significant at the 5% level) and the twelfth-grade boys scored 2 percentage points higher than the girls (not significant). The results of the comparisons between twelfth-grade boys and girls (when participation in mathematics courses is controlled) show little difference except for those students who had only one year of mathematics. For these twelfth-grade students having only one year of mathematics, the difference favoring the boys is significant at the 5% level for spatial visualization in two dimensions.

The 1975 study also provided an opportunity to compare interest preferences with those of 1960 for students in grades 9, 10, and 11. The 1975 group was slightly less interested than the 1960 group in the fields of business administration, fine arts and performing arts, and proprietors and sales workers. Interests were slightly higher for general labor and public and community service. Other changes included a greater interest by both boys and girls in the construction trades in 1975 than in 1960. With respect to technical jobs, such as computer operator, the girls' interest increased in 1975 and the boys' interest decreased as compared with their respective levels of interests in 1960.

The findings from these three projects may be summarized by emphasizing the need for a more concentrated study of mathematics, science, and social studies by the girls to improve the basic abilities necessary to perform effectively in a wider range of career fields. The results also suggest that such studies will prepare these girls for generally more effective functioning in this complex scientific age.

In terms of trends in overall performance over the 15-year period covered by these surveys, the results are somewhat mixed. Certainly the three tests for which amount of time studied is most directly productive of higher scores for all students—Vocabulary, English, and Computation—show substantially lower scores for both boys and girls. This suggests that students are studying less. This finding is in fact borne out by the comparison of students' reports on how many hours they study a week. They report studying about half an hour less in 1970 than in 1960. The scores with the largest gains represent subjects that were not directly taught as a part of the traditional curriculum—creativity and abstract reasoning. Perhaps these gains can be attributed to the newer trends in educational programs, including educational television and increased student initiative and responsibility. In a clearly culturally and vocationally related area, in Mechanical Reasoning and Visualization in Three Dimensions, the girls are improving while the boys are scoring a little less well than before.

In two important fields, reading comprehension and mathematics, the changes over the 15-year period are close to negligible. The scores are down one or two percentile points in Reading Comprehension and up two or three percentile points in mathematics. It seems highly likely that the trends noted are, at least in part, due to a larger percentage of the age group remaining in school through the eleventh grade in 1975 as compared with 1960. The fact that the less able students were inclined to drop out more frequently in 1960 certainly accounts for some of the drop in some scores over this time period. The second factor underlying these changes is related to changes in the curricula taken by the students in 1960 and 1975. The relative stability and, in some cases, increases in the average scores for such important factors as Abstract Reasoning, Creativity, Mathematics, and Reading Comprehension test results, despite the definite decline in the average ability levels of secondary school students because of improved student retention, is very encouraging.

REFERENCES

American Institutes for Research. *Planning career goals*. Monterey, California: CTB/McGraw-Hill, 1975–1977.

Armstrong, J. M. *Achievement and participation of women in mathematics: An overview*. Denver, Colorado: Education Commission of the States, 1980.

Flanagan, J. C. Quality education for girls. *Women's Education*, 1966, *5*, 1–7.

Flanagan, J. C. Changes in school levels of achievement: Project TALENT ten- and fifteen-year retests. *Educational Researcher*, 1976, *5*(8), pp. 9–12.

Flanagan, J. C., Davis, F. B., Dailey, J. T., Shaycoft, M. F., Orr, D. B., Goldberg, I., & Neyman, C. A., Jr. *The American high school student*. Pittsburgh: University of Pittsburgh and American Institutes for Research, 1964.

Shaycoft, M. F. *The high school years: Growth in cognitive skills*. Pittsburgh: University of Pittsburgh and American Institutes for Research, 1967.

Thorndike, E.L., & Lorge, I. *The teacher's word book of 30,000 words*. New York: Teachers College Press, 1944.

Chapter 4

Functional Literacy and Writing: Some Cautions about Interpretation

Roy H. Forbes

INTRODUCTION

The performance of American students in functional literacy and writing assessments is reviewed in this chapter. Two assessments in each of these areas have been conducted by the National Assessment of Educational Progress (NAEP). Data collected, analyzed, and reported from these assessments will be used to illustrate changes in student performance over time, while considering the rise and fall of test scores.

Each assessment evoked certain issues of general concern in the testing community; these are, definition of the subject area and standards- or criterion-setting for test scores, and measures to be applied to raw data. Each of these issues has major implications for the interpretation of test results, whether results are for a local district, a state, or the nation.

Functional literacy and writing are examined separately. To some readers, this may seem an artificial separation because basic aspects of writing are integral to literacy. However, the NAEP project did not include writing tasks in the functional literacy assessments for reasons that will be made clear in this section. This does not mean that writing should not be considered when investigating functional literacy; it does point to problems of definitions and of standards setting.

Writing will be the topic of the second section. Writing skills are difficult to assess because of some problems associated with standards setting, but

51

The Rise and Fall of National Test Scores

Copyright © 1982 by Academic Press, Inc.
All rights of reproduction in any form reserved.
ISBN 0-12-068580-9

particularly because of measurement techniques. Writing skills range from correct spelling and punctuation to good grammar, from proper syntax to overall effectiveness of the communication. This section highlights the need for a clear understanding of the measures to be applied prior to data interpretation.

Appendix A, at the end of this chapter, contains descriptive information about the NAEP. This way, the data presented can be placed in a historical and operational context. This information is important for the proper utilization of the data.

FUNCTIONAL LITERACY

Of all 17-year-old students in America, 13% are functionally illiterate.

Of 17-year-old students attending schools in low-income areas in America, 22% are functionally illiterate.

The statements above are dramatic. But before one can use these data to support any cause or to launch a new movement in support of something, the following questions should be considered.

1. What definition of functional literacy was used as a basis for items in the assessment?
2. Do percentages of functional illiterates change if the cut-off score (the standard for acceptable responses) used to determine the reported figures is changed?
3. Do these percentages represent a decline or increase in the number of functional illiterates from one point in time to another?

Contrasts in Definition

Estimates of illiteracy in America depend on the basis of the definition used for designing a measurement of literacy. Generally, definitions of literacy can be described as falling within one of two major categories: process or behavior. In a recent paper prepared by NAEP for the National Institute of Education, Brown (1980) described these two categories of literacy:

> Those who define literacy as a *process* see it as an intention to make meaning; an attempt to understand information; or a complex of largely unconscious psychological, cognitive, and social activities, most of which are beyond the reach of traditional measurement tools.

Those who define it *behaviorally* list many levels:
1. The ability to read and write one's name.
2. The ability to read such materials as are critical to "survival" (i.e., legal documents, health and safety information, job notices, application forms) and to write sufficiently to fill out forms.
3. The ability to perform reading and writing tasks required for performing one's job satisfactorily.
4. The ability to read (with comprehension) a range of materials for a variety of purposes and to write a range of communications for a variety of purposes and audiences.
5. The ability to perform (3) and to perform fundamental mathematical computations and access resources such as libraries.
6. The ability to perform (4) and to act aggressively in behalf of one's rights and responsibilities as a citizen.
7. Mastery of fundamental processes of reading, writing, problem solving, computing, speaking and listening; and mastery of a core knowledge base in the sciences and humanities; and mastery of basic tools for study—sufficient to enable one to pursue any personal goals in this society [pp. 11-12].

Given contrasting definitions of literacy, estimates of illiteracy in America range from 1% to 50% of the adult population (Adult Performance Level Project, 1975; Copperman, 1978; Fisher, 1978; Harman, 1970; Harman & Hunter, 1979; Harris, 1970; Murphy, 1973; NAEP, 1976; U.S. Census, 1969, 1977; Vogt, 1973). If one defines illiterates as people who are 14 years old or older but have completed fewer than 5 years of schooling, the figure is 2.8%. If one defines illiterates as people who cannot perform certain reading and writing tasks considered "functional" by some group of "experts," the figure can be as high as 50%. And if one defines literacy as the ability to read demanding materials with good comprehension and to write clear prose that is suitable to particular audiences, then most Americans may be in hot water [Brown, 1980, pp. 10-11].

Against the backdrop of a variety of definitions of functional literacy, NAEP conducted surveys in 1974 and 1975 to obtain data on the status of and changes in functional literacy. The national Right-to-Read effort awarded a grant to the Education Commission of the States so that NAEP could assess the functional literacy level of American youth. The major goal of these assessments was to determine the extent of functional literacy among 17-year-olds attending public and private schools.

In these assessments, functional literacy was defined as *being able to perform tasks necessary to function in American society, such as reading newspapers, instructions and even drivers' license tests.* This particular definition does not perfectly match any of those listed earlier, but is most closely related to number 2 in the list of behaviorally oriented definitions.

This definition evolved from a unique set of circumstances. NAEP had conducted a reading assessment during the 1970–1971 school year, and data for 17-year-olds were collected during March–May of 1971. Because of some resource limitations, the 1974 and 1975 Right-to-Read assessments used questions from the 1970–1971 reading assessment for 17-year-olds. Reading specialists from the Right-to-Read staff selected those items (from the available pool of items) that measured functional literacy-type activities. This mode of selection is somewhat different from the one typically applied in a full-scale national assessment. In a typical full-scale assessment, items are developed by subject-area specialists on the basis of what students should be able to achieve. However, the mode of selecting items was consistent with the purposes of the Right-to-Read assessments of functional literacy.

Appendix B contains an overview of the classification scheme and a brief description of the items selected and the questions asked about them. Most of the actual test items have remained unreleased so that they can be included in future assessments of functional literacy and so that changes in students' performance can be measured. The reading assessment pool obviously did not include writing items, so the functional literacy assessments had no writing tasks.

As a consequence of this process, the items used in the functional literacy assessments are based on a limited definition of literacy. In turn, this definition of literacy must be considered when interpreting the results quoted at the beginning of this section.

Contrasts in Criteria

Following the selection process described above, the Right-to-Read Mini-Assessment of Functional Literacy (MAFL) was administered to a sample of 17-year-old[1] students in public and private schools in March–May of 1974 and 1975. The Right-to-Read staff wished to establish a criterion or standard to be met by students in this age group so that an evaluation of functional literacy could be attained. Three graded criteria were designed to establish cut-off scores on the assessment items prior to data analyses (NAEP, 1976):

1. The ideal standard of performance was that every 17-year-old student be able to answer every item correctly. Therefore, *the desired performance level* (DPL) for students at age 17 was 100% correct.
2. The desired performance level was reviewed with National Assessment staff, who considered many of the selected items difficult for even the better reader. As a result, the desired performance level was

---

[1] In 1974, 17-year-olds were defined as all persons born between Oct. 1, 1956 and Sept. 30, 1957; in 1975, 17-year-olds were defined as all persons born between Oct. 1, 1957 and Sept. 30, 1958.

adjusted after administering selected items to a small sample of students in the Denver metropolitan area who had achieved above the ninetieth percentile on nationally recognized reading tests. The performance level obtained by this small sample was used as the basis for determining the *highest expected level of performance* for the functional literacy items.

3. Finally, a *minimally adequate performance level* meant that a 17-year-old student must be able to answer at least 75% of the items in order for his or her functional reading performance to be considered minimally adequate. Any students not meeting this criterion may be considered functionally illiterate [p. xi].

As a result of establishing these three criteria or standards for performance, the data analyses allowed for the implementation of cut-off scores. Results were made available for each cut-off score for the national sample and for each of the defined population groups within the sample. Tables 4.1, 4.2, and 4.3 (NAEP, 1976, pp. 41, 45, 47) display the mean percentages of success on the items for each of the criterion scores for the two administrations of the items.

At the beginning of this discussion on functional literacy, two dramatic results from the Mini-Assessment of Functional Literacy were presented. Both of these results were based on the criterion described as minimally adequate performance (75% correct responses). A review of the data in Table 4.3, Percentage for 1974 and 1975, will illustrate the differences in performance levels for the nation and for the low-metro group (economically disadvantaged) at each of the criterion cut-off scores. If the criterion of 60% is used, then approximately 3% of the 17-year-olds in the nation would be considered functionally illiterate and approximately 9% of the 17-year-olds residing in low-metro areas would be considered functionally illiterate.

The act of criterion- or standard-setting to assist in interpreting and evaluating data is, at best, an arbitrary process. Differences between such standard-setting scores can cause major fluctuations in the percentages of students meeting the established standards. However, standards can be justified and defended if they are consistent with the purposes of the test or assessment. In the case of the MAFL, the standards can be justified on the basis that there is a need to identify certain broad trends of literacy in the population. Moreover, data are often presented without relation to a standard or a comparative "yardstick" and are therefore of limited usefulness because they defy interpretation. Due to the establishment of a three-point criterion, the functional literacy skills of 17-year-olds were cast in a more complete context. This makes steps toward remediation or other forms of intervention possible.

TABLE 4.1
The Mean Percentages of Success (Desired Performance Standard) on
the COMPLETE MAFL in 1974 and 1975, the Changes from 1974 to 1975,
Standard Errors of the Changes, and the Probabilities that the
Changes Are Due to Random Error

| Variable Group | Mean Percentages | | 1974–1975 | | |
	1974	1975	Change	SECH	Prob.
National	87.5	87.6	+0.1	0.4	0.697
Region					
Southeast	84.7	84.8	+0.1	1.2	0.920
West	86.7	87.8	+1.1	0.7	0.112
Central	89.4	89.2	-0.2	0.6	0.711
Northeast	87.9	88.1	+0.2	0.9	0.865
Sex					
Male	86.5	86.9	+0.4	0.5	0.509
Female	88.4	88.3	-0.1	0.4	0.857
Race					
Black	76.3	76.9	+0.6	1.6	0.741
White	89.3	89.4	+0.1	0.4	0.928
Parental education					
No high school	82.1	81.3	-0.8	1.4	0.567
Some high school	84.5	83.1	-1.4	1.1	0.211
Graduated high school	87.0	85.0	+0.5	0.5	0.317
Post-high school	90.1	90.3	+0.2	0.4	0.555
Size/type of community					
Low metro	81.9	83.0	+1.1	1.5	0.465
Extreme rural	87.5	85.3	-2.2	1.5	0.142
Small places	87.4	87.8	+0.4	0.7	0.610
Medium city	87.7	87.2	-0.5	1.0	0.617
Main big city	87.9	87.1	-0.8	1.8	0.308
Urban fringe	88.7	89.9	+1.2	1.0	0.238
High metro	91.8	90.4	-1.4	1.1	0.194

Changes in Functional Literacy

By using items from the 1971 reading assessment for the 1974 and 1975
MAFL, it was possible to make comparisons between national and group
performance in three different time periods. Figure 4.1 (NAEP, 1976, pp.
12–13) displays the mean percentages of success based on the criterion for
highest expected level of performance for the nation and for each of the
defined population groups in the sample. This mid-range criterion, highest
expected level of performance, suggests a heartening picture of improvements from one point in time to another. Notice that the national performance levels increased in 1974 and 1975. In addition, the performance

TABLE 4.2
The Mean ADJUSTED Percentages of Success (Highest Expected
Level of Performance) on the COMPLETE MAFL in 1974 and 1975
and the Changes from 1974 to 1975

| Variable Group | Mean Percentages | | Change |
	1974	1975	1974–1975
National	91.2	91.4	+0.2
Region			
Southeast	88.1	88.3	+0.2
West	90.3	91.6	+1.3
Central	93.3	93.0	−0.3
Northeast	91.6	91.8	+0.2
Sex			
Male	90.2	90.6	+0.4
Female	92.0	91.9	−0.1
Race			
Black	79.1	79.7	+0.6
White	93.2	93.2	0.0
Parental education			
No high school	85.4	84.5	−0.9
Some high school	87.9	86.3	−1.6
Graduated high school	90.7	91.2	+0.5
Post-high school	94.0	94.3	+0.3
Size/type of community			
Low metro	85.2	86.4	+1.2
Extreme rural	91.2	88.8	−2.4
Small places	91.1	91.5	+0.4
Medium city	91.2	90.9	−0.3
Main big city	91.9	90.8	−1.1
Urban fringe	92.4	93.7	+1.3
High metro	95.9	94.1	−1.8

of many groups increased in both 1974 and 1975 and several groups made substantial improvements by 1975 even though their gains in 1974 were not statistically significant. For example, the low-metro group increased in mean percentages of correct responses in 1974 although the increase was not statistically significant. However, the increase for low-metro from 1971 to 1975 is statistically significant. Also, notice that the extreme rural group increased in mean percentages of correct responses from 1971 to 1974, but did not perform at the same level in 1975. On the other hand, the gains or increases in mean percentages of correct responses among blacks were statistically significant from 1971 to 1974 and from 1971 to 1975. The trend of improvements noted in this table suggests the narrowing of certain gaps in performance between groups.

TABLE 4.3
Percentages of 17-Year-Old Students Who Answered at Least 90%,
75% (Minimally Adequate Performance Standard) and 60% of the
MAFL Exercises in 1974 and 1975; the Changes from 1974 to 1975;
Standard Errors of the Changes; and Probabilities that the
Changes Are Due to Random Error

| | 90% of MAFL Exercises | | | | |
| | Percentage | | 1974-1975 | | |
Variable Group	1974	1975	Change	SECH	Prob.
National	55.6	55.6	0.0	1.8	1.000
Region					
Southeast	46.0	47.7	+1.7	3.7	0.653
West	55.6	54.2	-1.4	3.0	0.638
Central	60.9	61.5	+0.6	2.8	0.834
Northeast	56.0	56.5	+0.5	4.3	0.904
Sex					
Male	51.6	52.9	+1.3	2.2	0.549
Female	59.3	58.0	-1.3	2.2	0.549
Race					
Black	21.1	17.3	-3.8	3.4	0.258
White	61.1	61.6	+0.5	1.8	0.787
Parental education					
No high school	33.7	31.2	-2.5	4.8	0.603
Some high school	40.9	38.5	-2.4	4.1	0.555
Graduated high school	55.5	53.2	-2.3	2.2	0.298
Post-high school	64.0	66.5	+2.5	2.3	0.276
Size/type of community					
Low metro	34.8	38.0	+3.2	5.9	0.589
Extreme rural	56.9	48.5	-8.4	6.1	0.168
Small places	52.9	55.0	+2.1	3.5	0.549
Medium city	58.1	55.7	-2.4	3.4	0.478
Main big city	56.4	52.4	-4.0	7.9	0.617
Urban fringe	61.6	63.4	+1.8	5.2	0.682
High metro	74.6	66.4	-8.2[a]	3.8	0.032

Summary

In an era filled with controversy about educational testing and the rise and fall of test scores, there is a need for caution in interpreting test results and scores. The MAFL, conducted by the National Assessment of Educational Progress for the Right-to-Read program, has been presented to illustrate how different definitions and criterion-setting affect the interpretation process.

The three questions raised at the beginning of this section were selected

TABLE 4.3 (Continued)

	75% of MAFL Exercises				60% of MAFL Exercises				
Percentage		1974-1975			Percentage		1974-1975		
1974	1975	Change	SECH	Prob.	1974	1975	Change	SECH	Prob.
86.8	87.4	+0.6	1.1	0.596	97.0	97.1	+0.1	0.5	0.841
79.3	80.0	+0.7	3.6	0.841	94.2	93.3.	-0.9	1.5	0.562
86.7	88.0	+1.3	2.0	0.516	95.8	97.8	+2.0	1.1	0.080
91.2	90.9	-0.3	1.6	0.849	99.2	98.8	-0.4	0.6	0.509
86.9	89.0	+2.1	2.2	0.337	98.1	97.5	-0.6	0.8	0.478
84.7	85.4	+0.7	1.4	0.624	96.5	96.1	-0.4	0.7	0.589
88.7	89.1	+0.4	1.4	0.772	97.5	97.9	+0.4	0.5	0.441
56.6	58.4	+2.8	5.0	0.575	86.6	85.3	-1.3	2.9	0.653
91.7	91.8	+0.1	0.9	0.912	98.6	98.8	+0.2	0.4	0.569
77.7	70.7	-7.0	4.6	0.124	92.9	94.6	+1.7	1.9	0.379
79.1	76.6	-2.5	3.6	0.484	96.2	93.7	-2.5	2.0	0.215
85.7	87.5	+1.8	1.5	0.234	96.3	97.6	+1.3	0.8	0.087
93.1	93.9	+0.8	1.0	0.435	99.0	98.7	-0.3	0.4	0.390
73.0	78.5	+5.5	4.2	0.194	93.2	91.4	-1.8	2.4	0.441
87.3	80.3	-7.0	3.9	0.073	97.8	94.3	-3.5[a]	1.7	0.036
86.5	88.5	+2.0	1.9	0.294	97.1	97.7	+0.6	0.9	0.483
87.5	85.5	-2.0	2.8	0.484	97.0	96.4	-0.6	1.1	0.589
79.9	85.9	+6.0	4.7	0.201	96.8	97.0	+0.2	1.5	0.897
91.3	95.0	+3.7	2.2	0.095	98.7	99.1	+0.4	0.9	0.660
95.4	92.6	-2.8	3.3	0.401	99.1	99.4	+0.3	0.8	0.711

[a]Significant: The probability is no greater than .05 that the change is due to random error.

to create a healthy distrust of the findings (i.e., "of all 17-year-old students, 13% are functionally illiterate," and "of 17-year-old students attending schools serving low-income areas, 22% are functionally illiterate"). These findings must be interpreted in the broader context of the definition applied to functional literacy and the criteria or standards of performance established by the test designers and users. Neither of the findings is true for all definitions of literacy and for all criteria used to measure functional

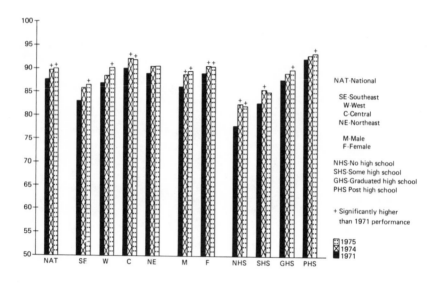

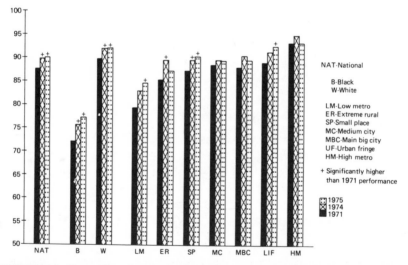

FIGURE 4.1. Mean adjusted percentages of success—highest expected level of performance. (Average of the percentages of 17-year-old students answering the individual exercises correctly adjusted to the percentages of superior readers answering the individual exercises correctly.)

literacy. In the case of some definitions and criteria, the findings are actually very encouraging; in the case of other definitions and criteria, the findings may be alarming.

MEASUREMENT OF WRITING SKILLS

This section contrasts two methods of measuring student performance on writing tasks. The National Assessment has conducted two writing assessments, the first during the 1969–1970 school year and the second during the 1973–1974 school year. A third writing assessment was conducted in 1978–1979, and results were published in 1980, but were unavailable at the time of this writing. Each assessment measured the writing skills of 9-, 13-, and 17-year-old students.

While both assessments consisted of a number of survey and multiple-choice questions and essay tasks, the range of tasks included in the 1973–1974 assessment was a bit more extensive than in the 1969–1970 assessment. Four reports are available on the 1973–1974 assessment of writing skills. The first contains the results of the 1973–1974 assessment and focuses on changes in the structure and mechanics of student composition since 1969. A second report focuses on information about expressive writing abilities. The third report focuses on persuasive and explanatory writing within the letter writing mode and the fourth concerns students' ability to make revisions in their own work.

In order to provide a frame of reference for the remainder of this discussion, one writing task administered to 13- and 17-year-olds and one task administered to 9-year-olds, followed by an abstract of the major results generated by these items, are reproduced below (NAEP, 1975).

Writing Task
AGES 13 AND 17

Everybody knows of something that is worth talking about. Maybe you know about a famous building like the Empire State Building in New York City or something like the Golden Gate Bridge in San Francisco. Or you might know a lot about the Mormon Tabernacle in Salt Lake City or the new sports stadium in Atlanta or St. Louis. Or you might be familiar with something from nature, like Niagara Falls, a gigantic wheat field, a grove of orange trees, or a part of a wide, muddy river like the Mississippi. There is probably something you can describe. Choose something you know about. It may be something from around where you live, or something you have seen while traveling, or something you have studied in school. Think about it for a while and then

write a description of what it looks like so that it could be recognized by someone who has read your description.
Name what you are describing and try to use your best writing.

Writing Task
AGE 9

Here is a picture [not shown here] of a kangaroo in Australia. Look at the picture for a while. What do you think is happening? Where do you suppose the kangaroo came from? Where do you think he is going? Look how high he jumps! Why do you suppose he is jumping over the fence?
Write a story about what is happening in the picture.

Results
17-YEAR-OLDS

1. There was an overall decline in the quality of the essays written for the second assessment: the mean holistic score dropped from 5.12 in 1969 to 4.85 in 1974, and the percentage of 17-year-olds writing papers ranked 4 or better declined from 85% to 78%.
2. Increases in awkwardness, run-on sentences and incoherent paragraphs most likely reduced the overall quality of the essays. Reduced coherence implies a diminishing of traditional organizational and transitional skills; awkwardness and the increase in run-ons suggest uneasiness with the conventions of written language. All of these changes point to a movement away from established writing conventions toward those of spoken discourse. More 17-year-olds may be writing as they speak.
3. In general, most of those aspects of writing generally called "mechanics" and stressed heavily in elementary and junior high school English classes (e.g., punctuation, capitalization, agreement, spelling, word usage and so on) are being handled adequately by the vast majority of students, and there is no evidence of deterioration in their use.
4. Good writers are as good as they were—i.e., have the same mean holistic score—and there may be a few more of them than there were in 1969.
5. Good writers are writing longer essays without losing coherence or increasing their error rates in such areas as punctuation, word choice, spelling, run-ons, fragments and so on.
6. Good essays contain about the same mixture of simple, compound and complex sentences and about the same proportion of sentences with phrases; they continue to contain only one spelling error in every 100 words.

7. Poor writers are worse than they were—i.e., have a lower mean holistic score—and there are more of them than there were in 1969.
8. Poor writers are writing shorter, less stylistically sophisticated essays but retaining about the same error rates—in effect, increasing their proportion of errors.
9. More poor essays are incoherent than were in 1969.
10. Poor writers are getting poorer, then, in those skills that are specific to written communication but seldom called for in conversation; that are acquired largely through broad reading and considerable rewriting; that are most seldom taught and, when taught, are most difficult to teach, especially to poor readers and people who have little use for printed communication.

13-YEAR-OLDS

1. There was an overall decline in the quality of the essays written for the second assessment: the mean holistic score dropped from 5.0 in 1969 to 4.7 in 1973, and the percentage of 13-year-olds writing papers ranked 4 or better declined from 79.6% to 76.6%.
2. The proportion of very good writers dropped from 19% to 13%.
3. There is a movement toward shorter, simpler expression. The essays were shorter in 1973 and contained fewer sentences with phrases. In other words, the students used less amplification and modification.
4. The vocabulary employed in 1973 was somewhat simpler.
5. There is a marked increase, particularly among males, in rambling prose, i.e., somewhat unfocused writing containing more run-on sentences and more awkwardness than was evident in 1969.
6. Most 13-year-olds commit at least one comma error in their essays.
7. More 13-year-olds are attempting to spell phonetically words they do not know.

9-YEAR-OLDS

1. The proportion of 9-year-olds writing papers ranked 4 or better rose from 51% in 1970 to 57% in 1974. The quality of the average essay written by a 9-year-old remained much the same between 1970 and 1974 and may actually have improved a bit.
2. The average paper in 1974 contains more complex sentences than the average paper in 1970, but it has lost paragraph coherence. In other words, 9-year-olds seem to be moving

toward more sophisticated writing, with mixed success. Trying to do more, they are risking more.
3. Most 9-year-old essays are free of run-on sentences, agreement errors, comma errors, period errors, word-choice errors and structure word errors.
4. Very few 9-year-olds write fully developed paragraphs focusing on a topic sentence, and the percentage is decreasing. The most rapid decrease is among the high-ranking papers [pp. 47–49, pp. 1–2].

The results above were obtained by "holistic" and "descriptive" scoring. Each essay was scored twice. First, the essays were scored holistically—a term derived from this method's emphasis on a reader's response to the *whole* essay rather than to such aspects of it as style, content, and mechanics. Readers were experienced English teachers, trained in the usual fashion by rating training papers—papers exemplifying each score point from 1 (the lowest quality score) to 8 (the highest)—until they had internalized the scoring system. Then each trained teacher read an essay and gave it a score from 1 to 8. Each essay was graded by two readers. If scores did not agree, the disagreement was resolved by a third reader.

The second scoring approach is descriptive, rather than evaluative. Another group of English teachers—all with considerable experience in grammar and linguistics—examined each essay according to a scoring guide. These readers coded each paragraph and sentence for its type (simple sentence, complex sentence with phrase, fragment, etc.) and coded each mechanical error (comma used when none required, no comma used when required, phonetic misspelling, fused sentence, etc.). The essays and the codes were keypunched and the results were tabulated by computer.

Whereas descriptive scoring focuses on some very specific mechanics and structural aspects of the written essays, the holistic scoring focuses more on the broad and general aspects of the essay as a whole. The descriptive score is rather straightforward and reliable. However, there are several questions that come to mind about holistic scoring:

1. What is a holistic score? What limitations are associated with holistic scoring of essays?
2. Are there better ways of scoring writing assessment data?

Some advantages occur as a result of holistic scoring. A reliable ranking of essays is possible once the scale points are defined and the readers are trained. The method is rapid—approximately one reading per minute including time spent organizing the papers. Therefore, holistic scoring can be a very practical method when a large number of papers require scoring.

However, National Assessment has found that there are some problems in explaining what the results of such scoring actually mean. For example, NAEP needs to report performance levels for particular writing skills and the rank ordering does not readily provide this information. In addition, NAEP requires a scoring system that can be replicated so that changes in performance can be measured over time. Moreover, other users of NAEP items and materials are interested in national results and in the replication of methods for their local and state assessments.

Ina Mullis (1980), of the NAEP staff, recently completed a paper pointing out some of the shortcomings of holistic scoring and suggested a new technique for scoring essays. Some of her remarks about the limitations of holistic scoring are reproduced below.

> 1. The relationship between the scorers' internalized criteria and the external or specified criteria is never described. Interpreters of results are given broad definitions and example papers. One never knows precisely why a paper received the rating it did. It is difficult to tell whether poorer papers all have good content and many mechanical errors, vice versa, mediocre content with some mechanical errors, syntactical problems, or what. Further analysis of the papers in each score point is necessary to collect this information.
>
> 2. Although rank ordering works well to separate the better papers from the poorer papers, holistic scoring is a relative process based on the quality of writing received. The terminology of top-half and bottom-half papers, although useful for training readers, has some inherent problems in that it tends to encourage a normal distribution. Generally, for every holistic scoring about 50% of the papers will be considered better papers and about 50% considered poorer papers. This does not necessarily mean that about 50% of the students can write well. In one sample the poor papers may represent competent writing and the better papers excellent writing. In another sample the poor papers may be nearly illiterate and the better papers just "less poor."
>
> 3. No attempt is made to state the standards for good or poor writing. "Goodness" is defined by the relationship of one paper to the rest of the papers in a particular set. Holistic scoring in and of itself does not provide information about how many respondents can write competently or adequately. Additional consideration must be given to the quality of the papers in each scale point to make such a determination. There is the risk that divisions suggested by the normal distribution of the papers will not be divisions suitable for the purpose of determining how many respondents can write with a specified degree of competence.
>
> 4. In practice, NAEP has found that even if the rank order of essays is replicated by a second group of readers, the location of the entire distribution of scores might not be reproduced satisfactorily. Because the criteria are so broad, it is difficult to standardize scoring sessions. Some papers are not clear examples of score points. In these cases readers must decide, for example, whether a paper is a 4 or a 5. Depending on the readers and/or the time of the reading, systematic differences can occur in these decisions. In one NAEP study, it was found that the *same* set of papers rescored holistically several years later had a significantly higher mean holistic score. Consequently, NAEP rescored all data from the first

Roy H. Forbes

writing assessment before calculating changes between the first and second writing assessments.

5. Independent evaluators will have trouble comparing their holistically scored data to NAEP holistically scored data. If the papers are scored differently from NAEP's, both sets (no matter how different) will tend to center around an average score. Therefore, it may appear that there is no difference. If there is a significant difference, it is not possible to determine whether it can be accounted for by differences in scoring sessions [pp. 3–5].

To illustrate the outcome of the holistic and descriptive approach, Tables 4.4, 4.5, 4.6, and 4.7 (NAEP, 1975, pp. 31, 32, 34, 36) show data from the 9-year-old "Kangaroo" writing task administered in 1970 and again in 1974. Similar data are available for 13- and 17-year-olds in NAEP's report, *Writing Mechanics, 1969–1974*. Some of these results were highlighted in the abstract shown earlier in this section.

Table 4.4 indicates a significant increase in average number of words per essay and per sentence from 1970 to 1974. However, Table 4.5 indicates a significant decrease in percentage of simple sentences with phrases per essay although the percentage of awkward sentences had also significantly decreased. Table 4.6 indicates that the average number of misspelled words had increased significantly, but the overall change is less than 1%. Table 4.7 shows that more essays by 9-year-olds are in the "middle to good" range in 1974 than in 1970.

TABLE 4.4
Average Essay Written by 9-Year-Olds, Counts

	1970	1974	Change
Average holistic score	3.8	4.1	0.3
Average number of words/essay	45.1	54.8	9.7[a]
Average number of sentences/essay	4.0	4.9	0.9[a]
Average number of paragraphs/essay	1.2	1.4	0.2[a]
Average number of puncuation marks	4.2	5.1	0.9[a]
Average number of letters/word	3.8	3.7	-0.1
Average number of words/sentence	13.9	15.0	1.1
Average number of words/paragraph	42.0	50.2	8.2[a]
Average number of sentences/paragraph	3.7	4.3	0.6[a]

[a] Differences that are statistically significant are indicated by small "a."

TABLE 4.5
Average Essay Written by 9-Year-Olds, Sentences

	1970		1974		Change	
	Average Number of Sentences/ Essay	Average Percent of Sentences/ Essay	Average Number of Sentences/ Essay	Average Percent of Sentences/ Essay	Average Number of Sentences/ Essay	Average Percent of Sentences/ Essay
Simple sentences	2.1	46	2.4	42	0.3	-4
Compound sentences	0.2	6	0.3	5	0.1[a]	-1
Complex sentences	1.0	25	1.3	27	0.3[a]	2
Run-ons	0.4	15	0.5	19	0.1[a]	4
Fragments (incorrect)	0.3	7	0.3	6	+[b]	-1
Sentences with phrases	2.4	60	2.7	55	0.3[a]	-5
Simple sentences with phrases	1.5	36	1.6	29	0.1	-7[a]
Complex sentences with phrases	0.7	20	0.9	21	0.2[a]	1
Awkward sentences	0.7	25	0.8	19	0.1	-6[a]

[a] Differences that are statistically significant are indicated by small "a."
[b] Plus signs equal rounded numbers less than .05.

TABLE 4.6
Average Essay Written by 9-Year-Olds, Spelling
and Word-Choice Errors

	1970	1974	Change
Average number of mispelled words	3.5	4.1	0.6[a]
Average percent of misspelled words	8.5	8.4	-0.1
Average number of word-choice errors	0.5	0.5	+[b]
Average percent of word-choice errors	1.1	1.0	-0.1

[a] Differences that are statistically significant are indicated by
small "a."

[b] Plus signs equal rounded percents less than 0.05.

This collection of tabular results, based on the holistic and descriptive
approach to scoring, indicates that 9-year-olds improved in certain writing
mechanics and structural skills from 1970 to 1974. However, the ability to
describe the more complicated aspects of writing performance is limited
because the criteria are not easily defined. Therefore, the raters' internal
criteria may intrude on the scoring process.

Due to the limitations of holistic scoring cited in the Mullis paper, the
National Assessment turned to the primary-trait system as a more appropriate measure of certain aspects of writing skills. This approach was
used for the first time with some of the data collected in the 1974 writing
assessment. The approach has since been developed more thoroughly and

TABLE 4.7
Changes in Essay Ratings, Age 9

	Lowest Score							Highest Score
	1	2	3	4	5	6	7	8
1970	3.6	19.2	25.8	19.6	12.7	12.7	4.3	2.0
1974	2.9	12.9	27.3	19.5	16.3	14.1	4.8	2.2

Percent of middle and good

1970	1974	Change
51.3	56.9	+5.6

has been used again to score writing tasks in the third writing assessment. However, the primary-trait system will not be the only measure used.

The major difference between holistic and primary-trait scoring is that the former tends to focus on the general quality of a written essay or letter while the latter focuses on the content and overall effectiveness of the communication. Although the structure and mechanics of a written piece may be perfect, the writing cannot be judged successful if the material does not accomplish its intended purpose. Four levels of success (inadequate, adequate, competent and excellent) are fully defined for the primary trait of each writing task, and responses are sorted into these categories by trained readers.

The 1974 writing assessment included a number of expressive, explanatory, and persuasive writing tasks. One such task developed for 17-year-olds involved writing an employment application letter. Writing a letter of application for a job involves a wide range of skills. Most important is the use of persuasive writing techniques and an awareness of the standard conventions of a routine business letter. An employment application item is reproduced here to illustrate how the primary-trait approach is applied (see Figure 4.2).

Among the questions addressed by the items are these: Does the writer understand the need to demonstrate interest in the job and provide data on qualifications? Can the writer articulate his or her past work experiences and make them specific and clearly relevant to the job being sought? Is the writer aware of the need to make his or her identity known and to state the specific purpose for writing? Are references included? Is an interview requested [NAEP, 1977, p. 15]?

All of these questions should be answered in the content of the letter. The extent to which these questions are answered in the letter determines the overall effectiveness of the communication and whether the purpose of the letter has been achieved.

Table 4.8 displays the percentages of responses in the primary-trait categories.

Tables 4.9, 4.10, and 4.11 (NAEP, 1977, pp. 18–19) show the results of the data analyses based on the specific aspects of information included in the letters. Each table provides a slightly different perspective on the content as well as on the quality of the letters written by 17-year-olds. As a result, a more complete picture of the writing performance data is available than would be possible with only the holistic scoring approach. If the employment application item were to be used to measure changes in performance, the primary-trait approach would assure more focused and detailed information because of the rigor and specificity of the criteria applied by raters from one time to the next.

Full-Time Job Application

Below are three ads from the Help Wanted section of
a newspaper. Read all three ads and choose which job
you would like best if you had to apply for one of
them.

·Help Wanted·

OFFICE HELPER: experience in light
typing and filing desirable but not necessary, must have 1 yr. high school math
and be able to get along with people,
$2.50/hr. to start.   Start now.   Good
working conditions.   Write to ACE
Company, P. O. Box 100, Columbia,
Texas 94082.

·Help Wanted·

SALESPERSON: some experience desirable but not necessary, must be willing to learn and be able to get along with
people. $2.50/hr. to start.   Job begins
now.   Write to ACE Shoestore, P. O.
Box 100, Columbia, Texas 94082.

·Help Wanted·

APPRENTICE MECHANIC: some experience working on cars desirable but
not necessary, must be willing to learn
and be able to get along with people.
$2.50/hr. to start.   Job begins now.
Write ACE Garage, P. O. Box 100,
Columbia, Texas 94082.

Write a letter applying for the job that you chose. Write
the letter as if you were actually trying to get the job.

FIGURE 4.2. Employment application item; a writing task for 17-year-olds.

TABLE 4.8
Percentages of Responses in Primary-Trait Categories for Employment-
Application Letter

Percentage	Trait
70.5	1 = The letter is in some sense incomplete. The writer is not identified and can't be contacted and/or the job being applied for is not described and/or qualifications for the job are not mentioned.
8.5	2 = The writer is identified and can be contacted, the job being applied for is described and at least one qualification for the job is mentioned.
0.7	3 = The writer is identified and can be contacted, the job being applied for is described and at least two general qualifications conerning experience, education or personal behavior are mentioned. The letter includes a greeting, closing and signature.

18.5   4 = The writer is identified and can be contacted,
            the job being applied for is described and at
            least two general qualifications concerning
            experience, education or personal behavior are
            mentioned. At least one qualification is
            specifically job-related. The letter includes
            a greeting, closing and signature.

1.8   Other = Students who did not respond to the exercise
            or who wrote illegibly.

TABLE 4.9
Percentages of 17-Year-Olds Giving Various Types of Information
on Full-Time Job Applications

Information	Age 17
Correctly gives job description	78
Describes qualifications	82
Provides references	9
Shows willingness to interview	16
Provides proper way to contact	36
Asks for more information	8
Gives reason to consider application	10
Gives personal behavior related to job	65

TABLE 4.10
Percentages of 17-Year-Olds Giving Various Qualifications
on Full-Time Job Applications

Qualifications	Age 17
1. EDUCATION AND TRAINING	
Gives level of school completed	13
Gives special related training	44
Gives general statement of education	2
2. EXPERIENCE	
Gives specific duties on previous job	3
Gives experience at specific job	46
Gives general statement of experience	16
3. PERSONAL QUALITIES	
Gives facts relevant to job	13
Gives behavioral traits relevant to job	65
Gives generalities not relevant to job	3

TABLE 4.11
Percentages of 17-Year-Olds Successfully Formatting the
Full-Time Job Applications

Format Category	Age 17
Return address	20
Date	32
Inside address	33
Appropriate greeting	92
Appropriate closing	85
Signature	85
All the above	8

CONCLUSION

One possible consequence, when certain aspects of test development and data analysis are overlooked, is inappropriate interpretation of data, fostering equally inappropriate solutions. Before broad generalizations can be drawn about student performance, issues such as definitions, criteria, and the limitations of measures should be considered. Performance data are far more complex than one may anticipate if confined to a single statistic.

When the functional literacy and writing data are considered in the broader context of other National Assessment findings, the following observation is suggested. The basic skills—functional literacy skills, writing mechanic skills, whole number computational skills, literal comprehension skills—are possessed by most students and do not appear to be declining. But higher order skills, the application of basic skills, do appear to be on the decline. However, the decline does not appear to be occurring in the upper level performance groups in some learning areas. The declines appear to be in the middle range of performance. Thus, a hypothesis requiring investigation is generated: Declines of higher order skills are most prevalent in the middle range performers. If this hypothesis is confirmed, then instead of the current hasty approach inherent in the back-to-basics movement, we can focus on and perhaps correct the problem.

APPENDIX A
NATIONAL ASSESSMENT: A SOURCE
OF DESCRIPTIVE EDUCATIONAL DATA

During the early 1960s when Francis Keppel was the Commissioner of Education, he became concerned with the fact that the Office of Education was not monitoring the nation's educational progress. This monitoring function was one of the Office's original responsibilities, but efforts had never been designed and implemented to meet this charge. Therefore, Mr. Keppel initiated a series of activities which resulted in the design and implementation of the National Assessment of Educational Progress.

At first, the NAEP was funded by the Carnegie Corporation, the Ford Foundation, and the Fund for the Advancement of Education. Today it is funded by the federal government through the National Institute of Education. Although federally funded, the program has always been controlled by a policy committee that is representative of the education community. Originally, the agreement describing NAEP's unique relationship between the government and the education community was spelled out in a memorandum of understanding signed by the Commissioner of Education and the Chairman of the Education Commission of the States, the education organization that has had the administrative responsibility of NAEP since 1969. In 1978, legislation formally placed the governance of the program with the Assessment Policy Committee. The committee has the authority to make all policy and operational decisions. It is the governing board for the NAEP.

The Assessment Policy Committee has adopted the following seven goals for the program:

1. *To detect the current status and report changes in the educational attainments of young Americans.* This represents National Assessment's original purpose; it is the overarching goal for the program upon which all other goals are built.
2. *To report long-term trends in the educational attainments of young Americans.* This recognizes the ongoing, long-range nature of the assessment program and the unique information it will provide. Assuming that the continuity and integrity of NAEP will be preserved, it will be possible to examine long-term trends as well as short-term changes and current statuses. The potential wealth of such information is truly awesome; its analog, the U.S. Census Bureau, is only beginning to tap fully the wealth of data it now possesses.
3. *To report assessment findings in the context of other data on educational and social conditions.* This goal recognizes that while the NAEP must remain true to its original purposes and not adopt an experimental research design, its data are best utilized by examining them in terms of other educational and social data. It will be possible to accomplish this through further analyses that link assessment data to other data bases and through minor alterations of data collection variables and techniques. This will ultimately result in greater ability to relate assessment data and in greater utility of the data for various levels of educational planners and decision-makers.

4. *To make the National Assessment data base available for research on educational issues, while protecting the privacy of both state and local units.* This goal recognizes the assessment's commitments both to open access to its data base and to honor the pledges of anonymity made to respondents and to state and local education agencies who have voluntarily participated in the assessment program. The present staff and policy committee members encourage the notion of secondary analyses of assessment data and attempt to facilitate such projects whenever and wherever possible. Only through greater utilization of assessment data (by both staff and external researchers) will NAEP's full potential be realized.

5. *To disseminate findings to the general public, to the federal government, and to other priority audiences.* This goal reflects the lessons that the present staff have learned over the years. To be useful, data and findings must be widely disseminated and suited to the desired audiences. Reports developed only for the research community do not suffice. This initially resulted in underutilization of the data and the view that the assessment program was a statistical maze. The current dissemination efforts, which gear materials to the specific audiences being served, have been quite successful and have helped raise the general acceptance and visibility of the assessment program to its present status.

6. *To advance assessment technology through an ongoing program of research and operation studies.* This goal acknowledges the program's commitment to greater efficiency (in terms of cost and data collection efforts), to continued refinement of measurement techniques, and to the development of new data collection and analysis techniques that enhance the usefulness of assessment results. The development of the primary-trait scoring system for writing is an example of staff efforts in refining and evolving measurement techniques. Development of a 4-year master sample and the subsequent decrease of sampling costs and burden on participating schools within that 4-year period provides an example of improved efficiency.

7. *To disseminate assessment methods and materials and to assist those who wish to apply them at national, state and local levels.* The final goal reflects NAEP's obligation to share its methods and materials with various education agencies. The program has proven to be immensely helpful to state and local education agencies and to the research community by releasing materials developed for regular assessment areas and probes (basic life skills, health, energy, and consumerism assessments). NAEP staff have also developed an excellent assistance program. To date, thirty-nine states have applied portions

of the assessment model, methods and materials to their own programs (Martin, 1979).

NAEP is designed to measure what 9-, 13-, and 17-year-old youths know and can do. Some attitudinal items are also included in each assessment. NAEP, in monitoring educational progress, has collected data for 11 years. Table 4A.1 lists the learning areas that have been assessed and provides information about the frequency of learning area assessments.

NAEP has administered probes in basic skills (1977) and consumerism (1978) for 17-year-olds and in health and energy (1977) for adults. In addition, the assessment program administered, analyzed, and reported the assessment of functional literacy of 17-year-old students under contract with the Right-to-Read program. Probes provide NAEP with the flexibility to address timely educational topics without disrupting its ongoing task of monitoring the nation's educational progress.

Following is a brief description of the methodology used by the NAEP in monitoring educational progress.

Objective Development

NAEP is objective-referenced. The objectives are developed using a consensus approach to determine what is important for students to know and be able to do.

Item Development

After objectives are developed, items are written. Items selected for the assessment are grouped into packages. Each package requires approximately 1 hour to administer. An individual student responds to only one package; therefore, a 13-year-old student participating in the assessment

TABLE 4.A.1
National Assessment Timetable, 1969-80 (Expressed in School Years)

Learning Area	Initial or Baseline Assessment	First Measurement of Change	Second Measurement of Change
Art	1974-75	1978-79	
Career and Occupational Development	1973-74		
Citizenship	1969-70	1975-76	
Literature	1970-71	1979-80	
Mathematics	1972-73	1977-78	
Music	1971-72	1978-79	
Reading	1970-71	1974-75	1979-80
Science	1969-70	1972-73	1976-77
Social Studies	1971-72	1975-76	
Writing	1969-70	1973-74	1978-79

may respond to only one-tenth of the items being used to assess that age group. Open-ended, short-answer, and essay questions are developed in addition to multiple-choice items. Items are field tested and reviewed for bias, validity, and reliability prior to being included in an assessment.

Data Collection

A multi-stage sample design, stratified by region of the country, size of community, and socioeconomic level, is used by the program. Schools selected in the sample are requested to voluntarily participate in the assessment. Data are collected by specially trained assessment administrators, employed by the Research Triangle Institute—the firm contracted by the Education Commission of the States—to draw the sample and collect data. Each item is responded to by approximately 2200–2600 individuals. Students' names are not recorded and commitments to the confidentiality of the data are made to the schools and school districts that participate in the assessment.

Analysis

Data are analyzed by the ECS assessment staff. Table 4A.2 lists the variables used in analyzing and reporting data. Results for groups, rather than for individuals, are calculated.

Reporting

Reports describing assessment results are published by NAEP. Both technical and lay reports are written to reach a broad and varied audience.

TABLE 4.A.2
National Assessment Reporting Variables

Variable	Group	Categories
Age	9-year-olds 13-year-olds 17-year-olds Adults (26–35 years)	
Region	Northeast	(Delaware, Connecticut, Maine, New Hampshire, Rhode Island, Vermont, District of Columbia, Maryland, Massachusets, New Jersey, Pennsylvania New York)
	Southeast	(Arkansas, Florida, Virginia, West Virginia, Alabama, Georgia, Kentucky, Louisiana, Mississippi, North Carolina, South Carolina, Tennessee)

Central	(Iowa, Kansas, Nebraska, North Dakota, South Dakota, Minnesota, Missouri, Illinois, Indiana, Michigan, Ohio, Wisconsin)
West	(Alaska, Hawaii, Idaho, Montana, Nevada, Wyoming, Utah, Arizona, Oregon, Colorado, New Mexico, Oklahoma, Texas, California, Washington)
Sex	Male, Female
Race	Black, White, Hispanic, and Other Other
Size and Type of Community	High Metropolitan, Low Metropolitan, Extreme Rural, Main Big City, Urban Fringe, Medium City, Small Places
Parents' Level of Education	No High School, Some High School, Graduate High School, Post High School

APPENDIX B
CLASSIFICATION AND DESCRIPTION OF THE ITEMS USED IN THE MINI-ASSESSMENT OF FUNCTIONAL LITERACY

CLASSIFICATION OF ITEMS

The types of formats of reading materials presented to the 17-year-old students in the MAFL were:

1. *Passages:* line-by-line narrative such as found in stories, poems, or newspaper and magazine articles.
2. *Graphic materials:*
   A. drawings or simple pictures, various signs including signs on doors and traffic signs, and a replica of a store coupon.
   B. charts, maps and graphs. The materials of this format could, in *one* sense, be considered reference materials, since a person consults these materials in search of specific information.
   C. forms such as an insurance policy statement, a check, a report card and a long distance telephone bill. Forms tend to be organized in a specific way to enhance the location of the materials they contain. These are also a feature of reference materials.

3. *Reference materials:* includes facsimiles of dictionaries, encyclopedias and telephone directories. These materials are organized to facilitate the location of specific information that a person is usually in search of when he consults a reference book.

The MAFL tested the 17-year-old's reading behaviors (skills) to see if they could:

1. *Understand word meanings:* Can the reader understand a specific word in context?
2. *Glean significant facts:* Can the reader locate or identify specific facts contained in the different kinds of reading materials?
3. *Comprehend main ideas and organization:* Can the reader identify the main idea or topic and understand how the writer organized facts to support it?
4. *Draw inferences:* Can the reader go beyond the information given and draw conclusions based on that information?
5. *Read critically:* Can the reader use his own thoughts and experiences to analyze, criticize, and evaluate; and then accept, modify or reject what the writer has said?

Labels attached to these categories of reading tasks can be misleading by implying greater difficulty than the actual tasks display. The functional literacy reading tasks required only a basic reading skill in all categories. For example, although comprehending main ideas and principles of organization generally implies a higher-order reading skill, the exercises included in this category represent a very low level of this skill. Four of the eight exercises merely required knowledge of the alphabetical organization of dictionaries, telephone books and encyclopedias; two asked for the main idea of very short passages (two and four lines); one asked which of four sentences did not belong with the others; and one asked with which fact a passage begins.

Description of Items

Because the items used in the MAFL are not in the public domain, they cannot be reproduced here. In lieu, a brief description of each item stem and the question(s) asked about it are presented.

1. List of words beginning with *FL* . . .
   Which of these words comes first in a dictionary?
2. Picture of four labeled doors that might be in a school—
   "Principal," "Nurse," "Cafeteria," "Library."
   Door where you might go for lunch?
3. Replica of automobile insurance policy statement.
   A. *What is the maximum amount for which this policy covers medical bills?*

   B. *What is the maximum amount this policy would pay*
   in case you injured another person in an automobile
   accident?
4. List of five pairs of last names beginning with "J . . ."
   You want to call Mr. Jones on the telephone. You will find his
   number between which names?
5. Facsimile of four common traffic signs.
   Which sign shows where you should ride your bicycle?
6. Facsimile of detail from a road map.
   A. *By car, is Northtown closer to Rice Lake than to Hope?*
   B. *Can you drive all the way from Northtown to Falls City on*
   Highway 71?
   C. *Is Hope the town closest to Centerville?*
   D. *Is Centerville farther west than Hope?*
   E. *Does Highway 20 run on the south side of Rice River?*
7. Facsimile of a check.
   A. *Who is to receive the money?*
   B. *What is the number of the check?*
   C. *For what reason would the bank probably NOT cash the*
   check?
8. Identify meaning of nonsense word as used in a passage.
9. Passage describing scarlet fever.
   Identify information given in first sentence.
10. Facsimile of a book-club membership application.
   A. *What shipping costs if you live in Canada?*
   B. *What money should you send with the order for books?*
   C. *How many additional books must you buy?*
11. Picture of door with sign hanging *on* it.
   Identify this fact as opposed to the signs hanging by, over, or near
   the door.
12. Identify which of four sentences does not belong with the other
   three.
13. Replica of a report card.
   Identify the reporting period covered.
14. Replica of "Help Wanted" advertisement.
   A. *Identify starting and quitting times.*
   B. *Identify minimum age.*
   C. *Identify salary.*
15. Picture of mushroom with parts named.
   Identify the parts named.
16. Replica of map taken from telephone directory showing service areas
   with prefixes and listing of which areas require toll charges.
   A. *Seven service areas are given. Identify which ones require a toll*
   charge and which ones do not.

     B. *If you placed a call to 533-0221, what area would you be calling?*

17. Identify sentence that tells the meaning of "I certainly won't miss that movie."

18. Replica of long distance telephone bill.
    A. *Where was the call made on February 14 from?*
    B. *Where was call made on February 14 to?*
    C. *What did call made on February 14 cost?*

19. Identify the word that best describes the tone of a sentence.

20. Facsimile of excerpt from 1040 income tax form instructions.
    Identify maximum standard deduction for married people filing separately.

21. Replica of an ad for preventing forest fires.
    A. *What is the name of the bear pictured?*
    B. *What is the purpose of the ad?*
    C. *What specific instruction does ad give to prevent forest fires?*

22. Replica of "Help Wanted" advertisement.
    A. *What company is offering the job?*
    B. *What kind of job is being offered?*
    C. *What are two qualifications for the job?*
    D. *What is the top salary after six weeks training?*
    E. *How should you apply for the job if you live locally?*
    F. *What do words "Equal Opportunity Employer" mean?*

23. Picture of four labeled doors that might be in a school—
    "Principal," "Nurse," "Cafeteria," "Library."
    Door where you would send a visitor who wants to see the person in charge of the school?

24. Recognize that a short passage tells how a man looks.

25. Replica of a report card.
    Identify subject area in which the student is improving his work.

26. Listing of comparable sizes for women's clothing in U.S.A., England, and Europe.
    A. *Size of shoes a woman would ask for in Europe if she wears 8 in U.S.A.*
    B. *If she wears a 38 sweater in the U.S.A., what size would she buy in Europe?*
    C. *If her wardrobe contains the following sizes: 34 blouse, 12 dress, 8 shoes, what sizes would she buy in Europe?*

27. Passage describing magic trick.
    A. *Identify the first thing to do to perform the trick.*
    B. *What do you do while standing in a dimly lit corner of the room?*

28. Picture of spines of a set of encyclopedias.
    Give number of volume where you would find out about George Washington.
29. Passage giving reasons for preferring dog over cat.
    Identify the number of reasons given.
30. Replica of traffic ticket.
    A. *Must a person appear at the Traffic Violation Bureau to plead "Not Guilty" to a violation?*
    B. *What would fine be for parking two hours in a one-hour zone?*
    C. *If a person found this ticket on his car Thursday, June 4, he would have until what date to pay his fine?*
31. Sentence distinguishing use of "ever" and "never."
32. Facsimile of billboard sign-horsepower.
    Identify where you would see this sign.
33. Passage describing a man's trip to Mt. Everest.
    Identify reason why man went to Mt. Everest.
34. Listing of telephone area codes and long distance information.
    A. *Identify number to call to obtain number in New York City.*
    B. *Identify number to call in Syracuse, N. Y.*
35. Facsimile of a label from a bug spray can, giving directions for use and list of bugs it kills.
    A. *Identify which bug probably would not be killed by spray.*
    B. *How far from surface should you hold the can?*
36. Facsimile of store coupon.
    A. *Identify "audience" the product will appeal to.*
    B. *Identify size(s) of product.*
    C. *Recognize terminal date for coupon use.*
    D. *Identify amount dealer will be paid for coupon.*
37. Replica of page from telephone directory giving information about long distance calls.
    A. *Recognize which rate would apply to station-to-station call made at 5:00 p.m. on a non-holiday Saturday.*
    B. *Recognize how a station-to-station call differs from a person-to-person call.*
38. Quotation of person commenting on TV programs.
    Identify the type of program the speaker prefers.
39. Facsimile of four common traffic signs.
    Identify which sign (for pedestrians) tells you what to do if you are walking.
40. List of five pairs of guide words from dictionary that begin with *O. . . .*
    Identify the pair between which you would find the word "optimal."

41. Facsimile of four common traffic signs.
    Identify sign to look for if you want to catch a bus.
42. Facsimile of portion of 1040 income tax form instructions.
    A. *Identify the marital status for tax purposes of a couple divorced during the tax year.*
    B. *Identify whether the couple could file a joint return the following year.*
    C. *Identify whether they could have filed a joint return for the year in which they were married.*
43. Passage
    Identify set of words that tell what the passage was mainly about.
44. Facsimile of contents listings from labels of two brands of dog food.
    Identify brand that contains more protein.

REFERENCES

Adult Performance Level Project. *Adult functional competency.* Austin, Texas: University of Texas, Division of Extension, 1975.

Brown, R. *Contributions of the National Assessment to understanding the problems of literacy and equity.* Denver, Colorado: National Assessment of Educational Progress, 1980.

Copperman, P. *The literacy hoax.* New York: William Morrow and Company, Inc., 1978.

Fisher, D.L. *Functional literacy and the schools.* Washington, D.C.: National Institute of Education, 1978.

Harman, D. Illiteracy: An overview. *Harvard Educational Review,* 1970, *40* (2), 226–243.

Harman, D., and Hunter, C.S.J. *Adult literacy in the United States: a report to the Ford Foundation.* New York: McGraw–Hill, 1979.

Harris, L., and Associates. *Survival literacy.* Conducted for the National Reading Council. New York: Louis Harris and Associates, 1970.

Martin, W. National Assessment of Educational Progress. In *New Directions for Testing and Measurement,* 1979 (2), 48–49.

Mullis, I. *Using the primary trait system for evaluating writing.* Denver, Colorado: National Assessment of Educational Progress, June 1980.

Murphy, R. *Adult functional reading study* (PR73-48). Princeton, New Jersey: Educational Testing Service, 1973.

National Assessment of Educational Progress. *Writing mechanics: 1969–1974* (Report No. 05-W-01). Denver, Colorado: Oct. 1975.

National Assessment of Educational Progress. *Functional literacy: Brief summary and highlights.* Denver, Colorado: 1976.

National Assessment of Educational Progress. *Explanatory and persuasive letter writing* (Report No. 05-W-03). Denver, Colorado: 1977.

U.S. Bureau of the Census: *Illiteracy in the United States: November 1969* (Population Reports, Series P-20, No. 217). Washington, D.C.: U.S. Government Printing Office, 1971.

U.S. Bureau of the Census. Personal communication, 1977.

Vogt, D. *Literacy among youths 12–17* (Vital and Health Statistics, Series 11–131). Washington, D.C.: U.S. Government Printing Office, 1973.

Chapter 5

Reading Trend Data in the United States: A Mandate for Caveats and Caution

Roger Farr
Leo Fay

INTRODUCTION

Because tests are constructed to measure specific goals at a specific time, no collection of test scores can adequately describe the literacy situation in the United States or literacy changes over time. Yet the right kind of achievement test data can, in part, attend to the need for improved schools by indicating whether contemporary students are scoring higher or lower. At the same time, an attempt can be made to confirm or deny the traditionally pessimistic view that students today are receiving an education that is inferior to the one that their parents received. In fact, one might assume that such criticism, which has been prevalent since at least the time of Ancient Greece, is based on some kind of achievement test data.

Yet that is not the case. Despite the fact that the most recent debate about reading (and other academic) achievement has been spurred by score declines, the current criticism of schools appears more than ever to be based on anecdotal evidence. This is particularly the case with the media, where carefully presented statistics may lose an audience, but a handful of selected interviews will "exemplify" while they entertain.

Criticism of our educational system has always been more opinionated than statistical, more anecdotal than research-oriented, more assumptive than conclusive. And when it has been based on data, it has failed to acknowledge the serious caveats that need to be considered. While the in-

The Rise and Fall of National Test Scores

Copyright © 1982 by Academic Press, Inc.
All rights of reproduction in any form reserved.
ISBN 0-12-068580-9

terest in achievement trend data is understandable enough to justify the need to seek the best information possible, using it can be as foolhardy as relying on anecdotes if one does not recognize its limitations. When it is reported that a group of current students scored better or worse than some previous group, the reaction to that "then and now" data should be prefaced with questions:

1. How well matched were the two samples in potentially crucial factors (e.g., educational experiences, age, intelligence, background, family income)? Are the two groups really representative of some larger population? How were they selected?
2. Was the same test used for both time periods? If not, how was it different? If it was the same, did it contain content that was out-dated for the later group tested? Does it still measure what is generally believed to be important in the domain covered?
3. How big is the score difference reported? What does the method of score reporting mean?

Depending on the answers to such questions, one can draw cautious conclusions that—*coupled to other kinds of information*—may help society understand trends in reading achievement.

Reading as a Whetstone

The long-running debate about the effectiveness of "today's education" was exemplified in Arthur E. Bestor's *Educational Wastelands* in 1953. He insisted that "educationists" had taken intellectual disciplines out of the educating process, and, as a result, children were not being taught how to think. More frequently, however, attacks on the development of language skills in the schools were less intellectual than Bestor's, merely citing examples of children's poor spelling and grammatical usage.[1]

Criticism of reading instruction was heated in 1955 with the appearance of Rudolph Flesch's *Why Johnny Can't Read*, which contended that phonics instruction in the schools had been replaced by a "look–say" method and that, as a result, the children of the nation were unable to read. Flesch's book had considerable impact and generated substantial lay and professional response.

In writing about and reviewing the book, most periodicals included critical responses. In *Newsweek* (March 21, 1955), William S. Gray stressed that there was more than one method to teach reading; in *Time* (June 20, 1955), Ruth Dunbar called the book a "hue and cry directed at a strawman." Flesch was subjected to analyses that pointed out numerous er-

---

[1] For example, an English teacher exemplifies language incompetence as spelling errors in the *Chicago Tribune*, February 16, 1962, p. 8.

rors in his book, that argued that he was writing about pronouncing—not understanding— words, and that insisted that phonics *were* being taught in conjunction with other methods. Several publications initiated lengthy series about how reading was being taught.[2] Such flare-ups in the debate continue yet today.[3]

The concern precipitated by Flesch boiled over with the advent of Sputnik in 1956. Although initially focused on science training, concern quickly expanded to education in general and to reading in particular. U.S. schools were compared with Russian schools in an attempt to explain how our nation lost the first lap of the race into space. Flesch's contention that we were ignoring the key answer—phonics—became the battle cry of the critics. An attempt in the *Saturday Review* to suggest that reading instruction is more complex than a phonics versus "look-say" dilemma earned a tart response from one reader: "There is a real war on in reading, and for the future well-being of American Education, it is important that the right side win.[4]

Arthur Trace exemplifies the impact of the space race on educational criticism. His *What Ivan Knows That Johnny Doesn't* (1961) insisted that, contrary to popular opinion, Russian schools did not neglect training in the humanities in favor of math and science. Rather, he insisted, they did a much better job than U.S. schools. In the *Saturday Evening Post* (May 27, 1961, p. 30) Trace compared the controlled vocabularies of American school reading texts to what he claimed were the much larger lexicons developed at the earliest ages in Russian pupils.

Trace's book and a collection edited by Charles C. Walcutt (1961) were typical of criticism in the early 1960s—they were not heavily supported with data. Oddly, there was no tendency in this debate to apply achievement trend data, which in those years would have shown marked gains in comparison to any previous periods.

A third great wave of concern and criticism has come with the reported decline in test scores—particularly on college entrance examinations—and it is, once again, highly focused on reading and reading-related areas.

The Relevance of Trend Data

Since score comparisons are intrinsic to the issue, reading trend information seems of particular significance. Unfortunately, there are limited quantities of such data overall, and what we have is so mixed in terms of

---

[2] For example, *Christian Science Monitor,* beginning October 7, 1955.

[3] Witness Flesch's re-emergence to revoice his argument in *Family Circle,* November 1, 1979.

[4] "But There Is No Peace," *Saturday Review,* April 21, 1962, p. 54. A response to comment in that periodical January 20, 1962.

subjects tested, tests used, time of year of testing, and other factors, that any generalizations drawn from the data are highly tenuous and subjective. For the sake of brevity, research that attempts to compare scores obtained at different times has been called "then and now" research.

There are also inherent caveats in then-and-now achievement research that would limit interpretations of it and conclusions drawn from it. These caveats would exist even if there were a large pool of reading achievement data. They are suggested by the questions above and will be discussed later in this chapter.

Finally, what then-and-now achievement data show about the successes or failures of our schools would not be absolutely clear even if there were sufficient achievement data and even if such data were not subject to inherent caveats. This is so because there is no assurance that any teaching theory, method, or practice is the single major factor in determining student achievement. It is acknowledged that many educational factors may affect achievement, and that they are closely related to other societal factors. These, too, will be discussed later. Once again, little research effort has been devoted to the study of how educational, economic, demographic, and other societal factors affect learning and performance.

But the few studies that have considered other factors (i.e., economic, demographic, societal) found these may be as important—if not more important—than educational factors in determining achievement outcomes. Surprisingly few then-and-now studies have considered, for example, something as basically pertinent as how population shifts have affected the make-up of groups of subjects whose reading performance scores are being compared over time.

It ought not be concluded from this relatively pessimistic introduction to a review of reading trends that it is appropriate to settle for selective use of the data that exist; to evaluate schools on general, unsubstantiated impressions; or to rely on the kind of highly limited anecdotal evidence so frequently presented in the media.

The immediate need is to attempt to consider the best reading trend evidence that exists. At the same time, such an analysis should consider the possible weight of all potentially relevant factors—both on existing as well as future data produced by then-and-now comparisons. Such a proposal suggests the possibility of a kind of cautious meta-analysis of all the available data, but it should not cloud the importance of recognizing differences in the data. These include the distinction between levels of comprehension and the differences between achievement and aptitude data; literacy data; functional competency information that may be very useful in planning curricula, but which has, as yet, no match in previous or subsequent data; general educational data that report on efforts to improve the equality—the comprehensiveness—of the educational system; and relevant then-and-now reading-related data other than achievement scores. There has been, in the relatively current debate, a tendency to mix such pieces of information without acknowledging their differences.

Part of the reason for careless use of existing reading trends data is the lack of appreciation for comprehensive analysis and synthesis. Such a project would involve an exhaustive search and a more diversified, objective effort than can be undertaken here. What this chapter will attempt to do is briefly review previous studies and analyses of reading achievement data and exemplify other trend data relevant to reading trends in the United States. The caveats related to conducting and interpreting then-and-now research will be discussed later in the chapter. These will incorporate some discussion of factors—other than instructional—which are obviously relevant to then-and-now reading achievement data.

THEN AND NOW READING ACHIEVEMENT

Using available reading achievement trend information, the following generalizations can be made:

1. The scope of the data in terms of periods covered, populations tested, and states or localities reporting trends is highly fragmented.
2. What data there are are very mixed in terms of reporting gains and losses.
3. What data there are cannot be reasonably used to substantiate claims that our schools are failing to teach reading.
4. In terms of longer time blocks, it appears that the reading achievement of the nation's students has been generally characterized by improvement.
5. In terms of the most basic competence in reading, schools have generally succeeded in developing literacy in the population.
6. In the lower elementary grades, improvement in reading achievement has been continuous throughout the history of education in the U.S.
7. There is some indication that the schools are not succeeding as one would wish in teaching the higher levels of reading comprehension.

To substantiate these generalizations, it is necessary to review a number of studies that have reported or analyzed then-and-now data.

Earliest Comparisons

In a thorough search of the literature, Farr, Tuinman, and Rowls (1974) reported that reading achievement for periods before the early 1900s can be deduced only from a handful of studies on general achievement. Assum-

Roger Farr and Leo Fay

ing that reading abililty could be justly deduced from students' abilities on achievement tests in subject matter areas that relate to or require reading, they reported finding six studies that produced test scores over a general period from 1845 to 1947. Five of the six reported gains on a total of 14 measures. A summary of these studies is given in Table 5.1.

It is questionable how well all the areas tested in these six studies—particularly arithmetic—reveal reading performance. These data are reviewed for three purposes. First, they do indicate gain in a very general sense; second, they show the scarcity of data for earlier periods; third, they demonstrate many of the problems endemic to then-and-now data. For example, the geographic spread for each study is very localized; the subject

TABLE 5.1
Earliest Studies on Achievement

Author (date of study)	Locale	Periods/Subjects Compared		Subject Areas	Results
Riley (1908)	Springfield Mass.	All (1846) ninth graders (245)	-to-	All (1905–1906) ninth graders (709)	Arithmetic Spelling Geography	Substantial mean gain
Caldwell & Courtis (1924)	Boston & U.S.	530 best 8th graders in Boston (1919)	-to-	12,000 8th graders all over U.S.	Seven academic areas (Local test)	Substantial average gain
Luther (1948)	Cleveland	35 8th graders (1848)	-to-	40 pupils (1947)	Six content areas lumped into one score (Local Boston test)	Slight gain
Fish (1930)	Boston	20 8th graders (1853)	-to-	200 8th graders (1929)	Arithmetic Grammar Geography (local Boston test)	Gains
Rogers (1946)	Chicago	16,000 6th graders (1923)	-to-	13,047 6th graders (1946)	Arithmetic (Woody-McCall Test)	Loss of .2 Grade Equivalent
Daughtry (1947)	Florida	4th, 5th, 6th graders in several counties	-to-	4th, 5th, 6th graders in two counties (unmatched)	Arithmetic Spelling Stanford Achievement Test (Form S)	Gains in all but 5th grade spelling

Source: Adapted from Tables One and Two in Farr. Tuinman, & Rowls, 1974.

samples for most are quite small; and the match of samples in most of the studies is not explained (e.g., in the Caldwell and Courtis [1924] study, it is not at all clear how 530 Boston students are a substitute for a U.S. sample or how the U.S. sample was drawn; in the Luther study, the size of the sample for Cleveland begs explanation of how and why the subjects were picked; and in the Daughtry study, the counties used in the later sample were not the same as those used in the earlier sample.

None of the studies considered how time might have changed the population of the locale and, thus, the samples. Several of these studies offer inadequate or no data analyses. Only two of them utilized a standardized test. One of the studies recognized an inherent problem in then-and-now comparisons and attempted to control for it. Caldwell and Courtis used the same test in 1919 as in 1845, but they used only 30 questions from it, eliminating questions they felt were invalid for the later sample.

Early Reading-Specific Comparisons

Twelve reading-specific comparison achievement studies are reviewed by Farr, Tuinman, and Rowls (1974). One of these studies reports losses, two report neither overall gains nor losses, and eight report gains. One reports losses until age differences are taken into account and then reports gains after that consideration.

These studies share some of the limitations exemplified by the six earliest and more general studies mentioned above. Statistical analyses in some studies are so inadequate or non-existent that the significance of gains or losses cannot be determined. Descriptions of subjects and the way they were selected is not clearly detailed in several, and the instruments used are not always adequately described.

On the other hand, several used large samples, and, as a group, these studies exhibit more awareness of factors outside the school that can affect reading achievement scores. One of the studies (Gates, 1961) is based on test-norming data and is, therefore, subject to the caveats relevant to such data and is discussed later in this chapter.

Study Showing Decline

Boss (1940) compared 8928 pupils from grades 1–8 tested in Saint Louis in 1916 with 1156 pupils from those grades in that city in 1938. The 1938 sample was matched to the 1916 sample using city-wide reading medians. The tests used covered silent and oral reading. She found the earlier sample superior in oral reading for grades 2–8 and "generally higher" on the silent reading in comprehension. The 1938 sample excelled at grades 2 and 3 in silent reading. She does not fully describe the instruments used nor does she note any age differences. Boss does point out, however, a key problem in then-and-now research by suggesting that the tests used were better matched to the earlier curriculum. Data from the Boss study are presented in Farr, Tuinman, and Rowls (1974, pp. 26-27).

Inconclusive Studies

In 1950, Burke and Anderson (1953) tested 216 pupils, grades 1-6, in Ottawa, Kansas, using the Metropolitan Achievement Tests. Their analysis of the pupils' backgrounds satisfied them that there were not significant differences between the 1950 subjects and the 162 pupils who took the test in 1939. They found no differences in reading at grades 2, 3, and 5. The 1939 group excelled in reading at grade 1, and the 1950 group at grade 4.

Krugman and Wrightstone (1945) utilized hundreds of thousands of scores for comparisons of New York City pupils tested in two time spans: 1935-1941 and 1944-1946. Farr, Tuinman and Rowls (FTR) review selected results, which are mixed; for example, 13,702 ninth graders tested in 1947 scored 4 months ahead of the national norm on the Nelson test, whereas 20,467 tested in 1938 had been but 1 month ahead of the national norm. In 1947, however, 21,252 eleventh graders tested with the Nelson Denny were 1 month above norm, whereas 29,319 eleventh graders were 2 months above in 1940. The effects of possible age differences and World War II (on the earlier sample) are not considered. The study is reported here as inconclusive, but it is highly possible that it suggests at least slight overall gains if all relevant factors are considered. This study's data for performance levels of grades 6, 7, and 8 on the *Stanford Reading Test* show that for the years 1935-1941 and 1944-1946, achievement was relatively stable. The data available are spotty within these years, but generally show that large samples scored from 0.1 to 2.0 months above the national norm. The data demonstrate no obvious trend (see FTR, 1974, p. 30).

Studies Showing Gains

A local study by Bradfield (1970) in rural California showed a non-significant gain on the Los Angeles Elementary Reading Test for a very small population of fifth graders in 1927-1928 and 1963-1964 comparisons (80 subjects, in total). An interesting aspect of this study is the relatively detailed attention given to the study of socioeconomic factors in order to rule out their possible effect.

Farr, Tuinman, and Rowls (1974) list three studies with gain results. Woods (1935) tested sixth graders in 33 Los Angeles schools in 1933 with the New Stanford Tests in Reading (1923-1924). The tests had been used in the same schools in 1923. Woods reported a mean gain of 6 months grade equivalent. Worcester and Kline (1947) used the Monroe Standard Silent Reading Tests (1921) in 1947 on over 5000 students from grades 3–8 in Lincoln, Nebraska and compared their scores to those of over 5000 students in the same grades from the same city on the same test in 1921. Mean and median gains were reported ranging from 18 to 3 points, decreasing with higher grade levels. Davis and Morgan (1955) found a 2-month grade equivalent gain for grade 6 children in Santa Monica, California, in 1939 when comparing their reading test scores to 1927 sixth graders from that city.

An early study of broader geographic significance was conducted by Tiegs (1949), who collected data on the Stanford Achievement Tests and the Progressive Achievement Tests from 60 communities in the states of New York, Pennsylvania, Delaware, Wisconsin, Michigan, Oregon, and California. Based on 82,733 cases, he found an overall gain in reading comprehension of one month, but a one-tenth of one month loss in vocabulary. For another 97,540 subjects, he reports an overall gain in reading of 1.8 months. The value of Tiegs' information, however, is limited by the fact that he gives no indication of what time span was covered when these "overall" differences occurred.

An interesting study to compare to Tiegs', in terms of implications for determining reading trends, is one by Finch and Gillenwater (1949). They tried to determine a balance between the number of subjects used and the thoroughness in reporting relevant information. Finch and Gillenwater found only slight gains for a small sample in Springfield, Missouri: The scores of 198 sixth graders in 1948 on the Thorndike-McCall Reading Scale (Form 3) were compared to those of 144 sixth graders who took the same test in 1931 (a mean of 23.32 in 1948 and a mean of 22.54 in 1931). The standard deviation for the 1931 group was 4.02, while for the 1948 group it was 6.32. The small sample enabled these researchers to make an admirably thorough study of potentially related factors. They found, for one thing, that the 1931 subjects were 1.56 months older than the 1948 subjects. They report also on the number of transfer students involved in the samples, on fathers' occupations, and consider the potential impact of "test-wiseness" on both samples.

Fridian (1958) studied scores on the Gates Reading Tests for all pupils in grades 1–7 in a parochial school in Lafayette, Indiana (1940–1956). Although the actual number of subjects is not reported, gains were found on all seven of the test's measures for all grades but the sixth. The mean gain across the seven grades was .523 grade equivalent months; the median gain was .60 (see FTR, 1974, p. 25).

Miller and Lanton (1956) compared scores on nationally standardized tests, using a total of 1828 subjects in grades 3–5 in Evanston, Illinois, for

20-year spans between the 1930s and 1950s. The New Stanford Achievement Tests were used in grade 4 and the Metropolitan Achievement Tests (MAT) in grades 3 and 5. Gains (in grade equivalent months) on the reading comprehension, paragraph meaning, and vocabulary subsections of the two tests ranged from .22 to .6. (See the February 1956 issue of *Elementary English,* Vol. 33, pp. 91–97; or Farr, Tuinman, & Rowls, 1974, p. 24.) Interestingly, Miller and Lanton conducted an examination of demographic and economic data in the community to determine that the population had remained stable between the time periods and that subject groups from both the 1930s and 1950s represented a "cross section" of racial, social, and economic factors.

The Gates Study

Noting the qualifying caveats associated with norming data (to be discussed later in this chapter), the Gates (1961) study is included in this section because it so clearly illustrates the importance of considering age differences. In renorming the Gates Reading Test in 1957, the author used careful sampling procedures to select 31,000 subjects he considered "geographically, economically, intellectually, and educationally representative of the United States at large." Their scores were compared to those of 107,000 students used in norming the 1937 test. For grades 2–6, the 1937 children excelled. But noting a striking difference in ages between the two groups at each grade level (from 4 to 6 months, depending on the grade level), Gates decided to compare results by chronological age. In that comparison, the 1957 pupils' scores were superior, leading Gates to conclude that the later sample had a reading ability a half a year advanced over children of the same age and intelligence 20 years before.

It should be noted, in reviewing the Gates study, that testmakers today are far less willing to endorse the use of their renorming data in the national then-and-now discussion—in spite of their highly scientific approach to conducting research. In-house analyses of norming data from different time periods are frequently made by publishers, but are not usually released for publication. This is not because such information reveals startling declines, but is (*a*) because of the testmaker's respect for the appropriate use of their data, which were not collected for achievement comparisons; and (*b*) because of the very serious caveats to using norm data in depicting trends. These are detailed later in this chapter, but to illustrate, the following is part of a response letter from the test department of one publisher to an inquiry about its norming data:[5]

---

[5] This quotation is taken from an in-house Psychological Corporation communication, with permission.

The difficulties of insuring comparability of pupil populations grade for grade, of controlling the consequences of changes in grade placement of topics, and of varied objectives, and problems of comparability of results yielded by different measuring instruments, make it extraordinarily difficult to obtain clear-cut comparisons of present and former achievement . . . . we have been reluctant to draw conclusions as to changes in achievement status from the data because of the many complicating factors that becloud the interpretation.

Summary of Early Reading-Specific Comparisons

The studies reviewed so far are disparate in terms of many key factors: the length of the time span between the scores compared; the actual time periods covered; the size, the urban–rural characteristics, and the geographic representation of the subject samples; the instruments used; the statistical analyses applied to the findings; and other research, societal, and educational considerations. Yet the scatter of some of these factors is what recommends some attempt at meta-analysis. Generally, the gains reported outweigh losses and suggest that over the broader span of time one can determine that the reading achievement scores of students in the United States increased.

This analysis is supported by all four of the summaries of reading achievement trend studies that were identified by Farr, Tuinman, and Rowls (1974). Witty and Coomer (1951) review seven studies and report on a survey of educators in 1947 (and on another in 1959). They conclude that "instruction today is as successful as it was at any period in the past [p. 457]."

Reports by the National Education Association, one on the three R's (1951) and another on general achievement (1952), review and table data to those dates to support the effectiveness of instruction. Gerberich (1952) reviews seven studies to argue that schools then were "not less effective" than in years before.

The then-and-now achievement studies that have been discussed are summarized in Table 5.2 (excepting Gates).

More Recent Reading Achievement Comparisons

Several of the more current then-and-now reading achievement studies show reading gains for the contemporary students. One of these has a limited subject pool—183 sixth graders in Ohio—whose scores on a short paragraph comprehension test that measured both literal and inferential comprehension were compared with sixth graders in the community who took the test in 1917 (Wray, 1978). Also, a five-word vocabulary test given

TABLE 5.2
Summary of 11 Reading-Specific Then and Now Comparisons

Study	Years Compared	Sites of Samples	Test(s) Used	Number of Samples	Gains or Losses	Grade(s) Tested
Boss (1940)	1916–1938	St. Louis	Unnamed test of silent and oral reading	(1916) – 3,923 (1938) – 1,856	Raw score losses	1 – 8
Worcester & Kline (1947)	1921–1947	Lincoln, Nebraska	Monroe Standard Silent Reading Tests, 1921	(1921) – 5,690 (1947) – 5,106	Marked mean and median score gains	3 – 8
Woods (1935)	1923–1933	Los Angeles	New Stanford Test in Reading, 1923–24	All grade 6 pupils from same 33 schools	6-month mean GE Gain	6
Bradfield (1970)	1927–28 to 1963–64	Rural California	Los Angeles Elementary Reading Test, Form 1	(1927–28) – 38 (1963–64) – 51	Non-signifi-cant mean raw score gain	5
Davis & Morgan (1955)	1927–1939	Santa Monica, California	Unnamed identical tests	All grade 6 pupils from city	2-month GE gain	6
Finch & Gillenwater (1949)	1931–1948	Springfield Maine	Thorndike-McCall Form 3, 1931	1931 – 144 pupils from 7 schools 1948 – 198 pupils from same 6 schools	Slight mean standard score gain	6
Miller & Lanton (1956)	1932–1953	Evanston, Illinois	New Stanford grades 4, 8; Metropolitan Achievement Tests, grades 3,5 (varying forms)	Total of all samples – 1,828	Slight mean GE gains at grades 3,5; marked median GE gain at grade 4; marked mean GE gain at grade 8	4:1932–52 3,5:1934–53 8:1933–54

TABLE 5.2 (Continued)

Study	Years Compared	Sites of Samples	Test(s) Used	Number of Samples	Gains or Losses	Grades Tested
Krugman & Wrightstone (1945)	1935-41 to 1944-47	New York city	Stanford; grades 6-8; Nelson, Nelson-Denny; grades 9, 11	(1935-41) - 290,000 Total (1944-46) - 242,000 Total	Non-signifi-cant gains	6 - 9, 11
Burke & Anderson (1953)	1939-1950	Ottawa, Kansas	Metropolitan Achievement Test, 1939	(1939) - 162 of 300 (1950) - 216 of 300	No signifi-cant mean GE differences in reading for grades 2,3,5; losses in grade 1; gains in grade 4; grade 6 results unreported	1 - 6
Fridian (1958)	1940-1956	Lafayette, Indiana	Gates Reading Tests, 1940	All pupils in one parochial school	Marked gains in mean GE, except for marked loss at grade 6	1 - 7
Tiegs (1949)	"Before and after 1945"	Sixty com-munities in New York, Penn-sylvania, Delaware, Wisconsin, Michigan, Oregon and California	Stanford Achievement Test and Progressive Achievement Tests	230,000 Total	"Slight, probably non-significant gain" in mean GE	4 - 11

Source: Farr, et al, 1974

to eighth graders in 1919 was administered to 583 eighth graders in 1978. The 1919 eighth graders did better than the 1978 sample, but the difference was not statistically significant. On the comprehension test, the 1978 sixth graders scored significantly higher than the 1917 students.

An Indiana Study

A study of reading achievement in Indiana (Farr, Fay, & Negley, 1978) compared the performance of nearly 8000 sixth graders and as many tenth graders in 1976 against a larger sample representing about one-fourth of the state's students in those grades in 1944–1945. The 1976 samples were drawn on a random design stratified to represent all geographic areas of the state and urban, suburban, small town, and rural schools in balance with the state's population. The earlier sample had demonstrated an effectively scattered representation. The same editions and forms of the Iowa Silent Reading Tests (ISRT) used in the 1940s were reprinted and used again in 1976.

The ISRT given to the sixth graders produced eight subskill scores and a total median score. In T-score and grade equivalent comparisons, the 1944–1945 sixth graders in Indiana outscored the later sample on three subtests and on total score. The 1976 sixth graders excelled on four of the subtests. It was found that the earlier sample were 10 months older, on average, than the current sample. In an age comparison, the 1976 sixth graders markedly outscored the earlier students on every subtest and on total score. The sixth-grade data are given in Table 5.3.

The ISRT used in the Indiana study to test tenth graders yielded nine subtest scores and a total median score. T-Score and grade equivalent comparisons showed the 1944–1945 sophomores outscoring the 1976 sample on five subtests and total score, while the latter sample excelled on four subtests. The 1976 sophomores, however, were found to be an average of 14 months younger than the students in that grade in 1944–1945. In an age comparison, the 1976 sample markedly outscored the earlier sophomores on total score and on all subtests but one (where there was no difference). The tenth-grade data are given in Table 5.4. It should be noted for later reference that for both sixth and tenth graders, age-adjusted gains were smaller the more global the comprehension required: word comprehension exceeded sentence comprehension, and sentence comprehension exceeded paragraph comprehension. The Use of Index subtest gains were largest for both sixth and tenth graders in 1976, and Directed Reading gains were high for both. Coupled to other data, this tendency to excel at the more specific reading skills may have some significance.

Several factors were examined and acknowledged as favoring the earlier students. Content analysis suggested that the tests were written for the

TABLE 5.3
T-Score[a] Comparisons of Achievement of Indiana Sixth Graders, 1944-45 and 1976, on the Iowa Silent Reading Tests (New Edition) -- Elementary Test: Form BM (Revised), 1943 -- Unadjusted and Adjusted for 10-Month Age Differences Between the Two Samples

Test	Unadjusted			Adjusted	
	Mean T-Score Achieved by 1944-45 Students	Mean T-Score Achieved by 1976 Students	1976 Gains or Losses	Mean T-Score Achieved by 1976 Students	1976 Gains or Losses
1. Rate	47.4	46.9	-0.5	48.9	+1.5
2. Comprehension	45.3	45.3	0.0	48.8	+3.5
3. Directed Reading	47.3	48.2	+0.9	52.3	+5.0
4. Word Meaning	45.1	46.3	+1.2	50.6	+5.5
5. Paragraph Comprehension	46.9	45.2	-1.7	48.9	+2.0
6. Sentence Meaning	47.2	47.2	0.0	51.1	+3.9
7. Alphabetizing	49.9	52.4	+2.5	54.2	+4.3
8. Use of Index	47.0	48.8	+1.8	52.9	+5.9
Total Median Score	46.2	45.4	-0.8	50.0	+3.8

[a]T-Scores have a mean of 50 and a standard deviation of 10.

Source: Farr, Fay, Negley, 1978.

values, content, and curriculum of that earlier period. Contemporary students are much less experienced with the rigid time constraints on tests such as the ISRT, since modern tests are timed more leniently to assure that about 95% of the test takers will finish. The 1976 sophomore sample included a percentage of poorer performers who would have dropped out of school before their sophomore year in 1944-1945—particularly in the light of opportunities for high salaries that existed in war industry in Indiana at that time. A look at demographic data shows that the minority population, which includes students for whom English is a second language, had increased nearly 2000% between the two time periods.

One factor noted favored the 1976 sixth-grade students. A survey accompanying each test administration revealed that the amount of time teachers reported devoting to reading instruction increased by 50% between the two

Roger Farr and Leo Fay

TABLE 5.4
T-Score[a] Comparisons of Achievement of Indiana Tenth Graders, 1944-45 and 1976, on the Iowa Silent Reading Tests (New Edition) -- Advanced Test: Form BM (Revised), 1939 -- Unadjusted and Adjusted for 14-Month Age Differences Between the Two Samples

Test	Unadjusted			Adjusted	
	Mean T-Score Achieved by 1944-45 Students	Mean T-Score Achieved by 1976 Students	1976 Gains or Losses	Mean T-Score Achieved by 1976 Students	1976 Gains or Losses
1. Rate	49.1	50.4	+1.3	50.9	+1.8
2. Comprehension	47.3	46.5	-0.8	50.0	+2.7
3. Directed Reading	46.7	47.8	+1.1	50.0	+3.3
4. Poetry Comprehension	48.6	47.6	-1.0	49.0	+0.4
5. Word Meaning	47.3	46.8	-0.5	50.0	+2.7
6. Sentence Meaning	45.8	43.2	-2.6	49.5	+3.7
7. Paragraph Comprehension	47.4	45.3	-2.1	47.4	0.0
8. Use of Index	47.5	50.0	+2.5	52.9	+5.4
9. Selection of Key Words	48.9	49.5	+0.6	51.1	+2.2
Total Median Score	48.0	47.4	-0.6	50.0	+2.0

[a]T-Scores have a mean of 50 and a standard deviation of 10.

Source: Farr, Fay, Negley, 1978.

periods: Such instruction averaged 37 minutes in 1944-1945 and 58 minutes in 1976.

In addition, the Indiana study suggests the importance of considering a host of other factors—such as the influence of television—that were not carefully examined at the time of the study.

NAEP Data[6]

The National Assessment of Educational Progress (NAEP) is one of the broadest evaluations of reading performance. It has been conducted in

[6] See also Chapter 4 in this volume (Eds.).

1971 and 1975 and is again being administered at the time of this writing. The NAEP random samples are painstakingly designed to represent the whole of the nation's children at the ages of 9, 13, and 17.

The NAEP reading test is an attempt to assess basic literacy. Its literal, inferential, and reference skill items are written using everyday reading tasks with practical significance. Thus it is a reasonable measure of *literacy* and should realistically correspond to the minimum competency expectations that are dictating tests of this nature evolving throughout the United States.

The 9-year-olds in 1975 performed significantly better on all three measures of the NAEP reading test than did the 9-year-olds in 1971, and black 9-year-olds improved even more dramatically than the age group did as a whole.

The 13- and 17-year-olds in 1975 improved slightly on literal comprehension and decreased slightly on inferential comprehension. The 17-year-olds also improved on reference skills in 1975.

The NAEP data on reading are given for the nation as a whole and then by regions, sex, and race for each age group in Tables 5.5, 5.6, and 5.7. The performance differences between such factors recommend the importance of considering societal factors when looking for reading achievement trends.

Data from Metropolitan Areas

The following data are limited in terms of area, years, and grades covered and information reported, and they are offered here in the hope that they may contribute to more dependable trend pictures when coupled to future data.

Farr, Tuinman, and Rowls (1974) contacted 27 of the nation's largest metropolitan school districts and 73 smaller school districts and then circulated a questionnaire to get reading achievement trend data over a period of 10–20 years covering a large geographic area. The researchers pursued persistent follow-up procedures in the survey.

The 100 districts covered every state and different parts of highly populated states. The results of this effort were indicative of the limited data that appeared to be available and of the difficulty of getting school personnel to release data when there was an indication that data existed.

Of the 27 most populated areas, 17 responded and of those, 16 said they had conducted reading achievement tests in the past two decades. Of these, 7 answered the request for summary reports of the testing results (the city school districts of Los Angeles, New York, Houston, Detroit, Milwaukee, and those from the urban districts of Dade County, Florida; and Montgomery County, Maryland).

TABLE 5.5
Mean Percentages[a] and Standard Errors of Differences[b] for 9-Year-Old Students on Literal Comprehension, Inferential Comprehension, and Reference Skill Items for NAEP Reading Assessments, 1970-71, 1974-75

	Literal Comprehension			Inferential Comprehension			Reference Skills		
	Mean Percentages		SE	Mean Percentages		SE	Mean Percentages		SE
	1970-71	1974-75		1970-71	1974-75		1970-71	1974-75	
Nation	65.745	66.780	.620	60.476	61.370	.615	64.784	67.003[c]	.708
Region									
Southeast	60.347	63.784[c]	1.568	55.746	57.654	1.358	60.906	64.101	1.813
West	65.364	65.295	1.239	59.734	60.800	1.166	63.537	64.993	1.482
Central	68.326	68.932	1.007	62.965	63.570	1.207	67.952	69.896	1.258
Northeast	67.687	68.567	1.181	62.379	62.856	1.145	65.501	68.429[c]	1.339
Sex									
Male	63.319	64.459	.680	58.270	59.593	.704	62.209	64.350[c]	.885
Female	68.142	69.120	.695	62.653	63.141	.636	67.362	69.619[c]	.885
Race									
Black	51.527	56.287[c]	1.435	46.866	50.529[c]	1.361	49.443	56.470[c]	1.494
White	68.196	69.189	.569	62.869	63.850	.612	67.356	69.532[c]	.743

[a] The mean percentage for a set of items is the average percentage of students who responded correctly to the items.

[b] The standard error (SE) of the difference between a 1970-71 mean percentage and a 1974-75 mean percentage was computed by taking the square root of the sum of the squared 1970-71 standard error of the mean and the squared 1974-75 standard error of the mean.

[c] Indicates that the difference in mean percentages between the two assessments was at least two standard errors in magnitude. The probability that a difference this large would have occurred by chance is .05 or less.

Source: Tierney and Lapp, 1979.

Reprinted with permission of the authors and the International Reading Association

TABLE 5.6
Mean Percentages[a] and Standard Errors of Differences[b] for 13-Year-Old Students on Literal Comprehension, Inferential Comprehension, and Reference Skill Items for NAEP Reading Assessments, 1970-71, 1974-75

| | Literal Comprehension | | | Inferential Comprehension | | | Reference Skills | | |
| | Mean Percentages | | SE | Mean Percentages | | SE | Mean Percentages | | SE |
	1970-71	1974-75		1970-71	1974-75		1970-71	1974-75	
Nation	61.817	62.722	.676	56.065	55.279	.566	65.678	63.831	.847
Region									
Southeast	56.081	58.755[c]	1.315	52.435	52.767	1.170	59.819	59.098	1.667
West	61.010	62.001	1.092	55.478	54.075	.845	65.622	62.508	1.559
Central	64.485	65.613	1.232	58.032	57.451	1.203	69.174	66.770	1.386
Northeast	64.941	64.010	1.585	57.846	56.449	1.174	67.188	66.331	1.974
Sex									
Male	58.849	59.714	.761	53.987	53.003	.686	63.820	62.365	1.126
Female	64.776	65.717	.713	58.145	57.524	.629	67.647	65.297[c]	.902
Race									
Black	46.241	47.700	1.212	43.222	43.870	1.127	47.764	45.508	1.913
White	64.598	65.657	.616	58.372	57.560	.549	68.692	67.196	.759

[a] The mean percentage for a set of items is the average percentage of students who responded correctly to the items.

[b] The standard error (SE) of the difference between a 1970-71 mean percentage and a 1974-75 mean percentage was computed by taking the square root of the sum of the squared 1970-71 standard error of the mean and the squared 1974-75 standard error of the mean.

[c] Indicates that the difference in mean percentages between the two assessments was at least two standard errors in magnitude. The probability that a difference this large would have occurred by chance is .05 or less.

Source: Tierney and Lapp, 1979.
Reprinted with permission of the authors and the International Reading Association

TABLE 5.7

Mean Percentages[a] and Standard Errors of Differences[b] for 17-Year-Old Students on Literal Comprehension, Inferential Comprehension, and Reference Skill Items for NAEP Reading Assessment, 1970-71, 1974-75

	Literal Comprehension Mean Percentages		SE	Inferential Comprehension Mean Percentages		SE	Reference Skills Mean Percentages		SE
	1970-71	1974-75		1970-71	1974-75		1970-71	1974-75	
Nation	76.789	77.045	.494	64.238	63.210	.594	69.224	69.530	.858
Region									
Southeast	71.953	73.087	1.142	59.405	59.760	1.206	62.196	63.208	2.096
West	76.255	76.161	.995	63.324	62.047	1.140	68.628	68.347	1.879
Central	78.956	79.452	.720	66.356	65.207	.984	72.373	72.897	1.271
Northeast	78.525	78.314	1.042	66.290	64.779	1.378	71.547	71.743	1.676
Sex									
Male	74.834	74.859	.626	61.946	61.537	.635	68.860	68.609	1.015
Female	78.645	79.109	.504	66.425	64.805[c]	.753	69.581	70.411	.942
Race									
Black	61.626	63.310	1.408	46.956	47.308	1.178	45.365	47.725	1.967
White	78.857	79.632	.421	66.517	65.980	.529	72.310	73.304	.657

[a]The mean percentage for a set of items is the average percentage of students who responded correctly to the items.

[b]The standard error (SE) of the difference between a 1970-71 mean percentage and a 1974-75 mean percentage was computed by taking the square root of the sum of the squared 1970-71 standard error of the mean and the squared 1974-75 standard error of the mean.

[c]Indicates that the difference in mean percentages between the two assessments was at least two standard errors in magnitude. The probability that a difference this large would have occurred by chance is .05 or less.

Source: Tierney and Lapp, 1979.

Reprinted with permission of the authors and the International Reading Association

102

Of the 73 smaller districts, 25 out of 28 that responded had done testing, but only 12 had summarized their data, and only five of these would make the summaries available (Anchorage Borough, Alaska; Jonesboro, Arkansas; Duval County, Florida; Worcester, Massachusetts; and Tacoma, Washington).

Overall, the data from the 12 districts of the original 100 (seven large metropolitan and five smaller districts) that made summaries available represented urban, suburban, and rural areas in all major geographic areas of the country. Only six of these districts, however, furnished data covering a period of 4 or more years.

The conclusions from these six sets of test data can be drawn only with great caution. In general, the test scores which cover the years from the late 1950s to the late 1960s and early 1970s indicate a pattern of slightly increasing test scores. District A shows this pattern at grades 3–9 for the years 1962–1971 (see Table 5.8). District B shows a pattern of increasing test scores at grades 3, 4, and 5 (see Table 5.9); District C indicates an upward trend at grade 7 (Table 5.10); District D reveals upward trends at grades 5 and 6 (Table 5.11); District E scores increase at grades 3, 4, 5, 6, and 8 (Tables 5.12 and 5.13); District F indicates a fairly stable pattern at grades 4 and 6 (Table 5.14).

Decreasing test scores are also revealed in some of these districts. District C, which indicated increased scores at grade 7, indicates decreasing scores at grades 3, 4, and 5. District E, which indicated increasing grade scores through 1965, indicates decreasing scores from 1966 through 1972. District F indicates very slight decreases at grades 8, 10, and 12.

There does not seem to be any very strong pattern in these data. These data do not corroborate either the consistent increase at lower grade levels or the decrease at upper grade levels that are generally assumed by trend analysts. Also contradicted is an increase in the earlier years (1960–1965) and a decrease in the later years (1965–1970). Perhaps the most important conclusion that can be drawn from the test scores of these six school systems is that there has been relatively little change in the test scores. At a time when the school populations of most cities in this nation were undergoing the strains of budget problems, integration, and, in some cases, increasing student populations, the average reading performance of students seemed to be holding fairly steady.

While it is often appealing to draw grandiose conclusions, there are obvious limitations of the data from these six school systems:

1. The sample is very limited.
2. The data for grades and year spans are spotty within this limited sample.
3. The reports for the systems are fraught with factors that make

generalization risky (i.e., changes in tests and forms used, lack of
detail in norming tests and forms used and in explaining how samples
tested were selected, differences in reporting, such as separation of
performance by race).
4. Types of tests and scores used are mixed both across and within some
   districts, and the scores' actual relevance to reading ability is not
   always clear.
5. Gains and losses are subject to some fluctuation which tends to
   frustrate identification of trends.

The following eight tables give the data reported by the six districts.

Data from States

The data presented for states are from three efforts: (*a*) FTR (1974)
surveyed all 50 states for trend data. Only eight states reported data for
three or more years; (*b*) Fay conducted an unpublished study in 1978 and
acquired data from 24 states, but only a few of these provided data or com-

TABLE 5.8
Data from Testing in School District A: Iowa Tests of Basic Skills[a],
given in 1962-1971; Total Test Performance in Grade Equivalents

Year		Grade						
		3	4	5	6	7	8	9
1962	white	3.8	4.7	6.1	7.2	7.7	8.9	9.3
(Midyear)	black	2.9	3.3	4.2	5.4	6.2	6.5	7.4
1963	white	4.0	4.6	6.0	7.2	8.0	8.6	9.2
(Dec.)	black	2.8	3.9	4.4	5.2	6.1	7.0	6.9
1965	white	4.4	4.6	5.6	6.7	7.7	8.7	–
(May)	black	2.8	3.1	3.8	4.7	5.5	5.8	6.6
1967 (May)		–	4.9	–	6.6	–	8.3	–
1968 (May)		–	4.9	–	6.7	–	8.2	–
1969 (May)		–	4.9	–	6.7	–	8.2	–
1970 (May)		4.1	4.7	5.8	6.8	7.6	8.2	8.7
1971 (May)		4.1	4.9	5.8	6.7	7.6	8.5	8.7

[a]Data from the ITBS covers reading, arithmetic, language skills,
and work study skills. Therefore, specific inferences about
reading achievement are limited.

Source: Farr, et al, 1974.

TABLE 5.9
Reading Data in Median Grade Equivalents from Testing in School
District B: SRA[d] Achievement Series, 1959-1971

Year	Grade					
	3	4	5	6	7	8
1959[a]	3.2	4.1	5.0	6.7	7.3	8.6
1960	3.4	4.5	5.5	7.0	8.1	8.9
1961	3.5	4.5	5.4	7.2	8.1	9.1
1962	3.6	4.5	5.3	7.4	8.0	8.8
1963	3.7	4.7	5.2	7.1	8.2	9.1
1964[b]	3.6	4.8	6.0	7.1	-	-
1965	3.6	4.6	5.5	7.0	-	-
1966	-	-	5.8	7.0	-	-
1967	2.9	-	5.6	-	-	-
1968	3.1	-	5.3	-	-	-
1969	-	-	5.6	-	-	-
1970	3.3	4.1	5.1	6.0	-	-
1971[c]	4.1	4.9	5.7	6.6	-	-

[a] Testing from 1959-1970 was done in September and October.
[b] A renormed version of the test was used beginning with this date.
[c] In 1971, testing was done in April.
[d] Science Research Associates

Source: Farr, et al, 1974.

ments on performance over a time span; and (c) data from several states
have been sent to the Reading Research Center at Indiana University or acquired from published sources or state education department releases. In a
recent attempt to identify states with more data than this review includes,
24 states with the potential to update or enlarge on this information or
with probable on-going assessments in reading were contacted. Fifteen of
these attended a session on uncovering and developing reading trend data,
but only two of those supplied additional information, and even that was
limited. Overall, the limits of the data in this section are presented with the
same caution as that in the previous section: Scattered, weak data can be
used to draw only weak inferences.

TABLE 5.10

Data in Median Percentiles from Testing in School District C:
Grades 3, 5, 7: Lorge-Thorndike and Iowa Tests of Basic Skills:
Grades 9, 11: Test of Academic Progress and Lorge-Thorndike,
1966-1972.

Year	Lorge-Thorndike - Verbal				
	3	5	7	9	11
1966	71[a]	71	69	69	83[c]
1967	71	69	69	69	73
1968	71	67	71	69	75
1969	71	67	71	69	75
1970	71	69	69	67	81[d]
1971	71[b]	67[b]	67[b]	65	75
1972	71	71	69	71	–

	Iowa Tests of Basic Skills								
	Verbal			Reading			Composite[a]		
	3	5	7	3	5	7	3	5	7
1966	65	70	69	68	64	59	76	71	66
1967	65	67	69	66	62	59	76	69	64
1968	65	70	65	66	62	58	73	69	64
1969	65	65	62	66	60	56	70	67	62
1970	65	67	67	63	60	59	73	67	62
1971	62	67	67	63	57	58	73	67	62
1972	69	65	68	63	62	63	73	70	70

	Test of Academic Progress			
	Reading		Composite[a]	
	9	11	9	11
1966	65	66	n.a.	67
1967	65	59	69	66
1968	65	59	65	66
1969	65	56	69	66
1970	65	59	65	66
1971	61	52	65	62
1972	61	–	65	–

[a] Includes non-verbal scores.

[b] 1972 scores derived from the Cognitive Abilities Test.

[c] Grade 12 tested in 1966.

[d] Data based on 5 schools only.

Source: Farr, et al, 1974.

TABLE 5.11

Data in Mean Grade Equivalents from Testing in School District D:
Grades 5 and 6: California Achievement Test Battery, 1954–1961

Year	Reading			Language		
	Mean G.E.	S.D.	# of S's	Mean G.E.	S.D.	# of S's
			Fifth Grade			
1954	5.217	1.17	2046	5.63	.98	2039
1955	5.30	1.41	1887	6.14	1.17	1906
1956	5.256	1.33	2188	5.621	1.03	2183
1957	5.44	1.37	2711	5.70	.99	2718
1958	5.25	1.26	2545	5.75	1.03	2554
1959[a]	5.99	1.43	2497	5.78	1.39	2495
1960[a]	5.92	1.44	2447	6.00	1.39	2444
			Sixth Grade			
1955	6.58	1.32	2140	6.54	1.01	2138
1956	6.55	1.43	1996	7.07	1.19	1987
1957	6.54	1.38	1858	6.56	1.10	1852
1958	6.69	1.41	2122	6.61	1.06	2117
1959[a]	7.31	1.36	2745	6.96	1.38	2737
1960[a]	7.32	1.36	2461	7.07	1.34	2465
1961[a]	7.33	1.40	2434	7.40	1.49	2432

[a]New form of California Achievement Test Battery used.

Source: Farr, et al, 1974.

The only clear generalization that can be derived from the data on the states is that drawing any conclusions on reading trends from it is very difficult, if not impossible. The scores reported—both within and across states—are, at best, spotty in terms of grades covered, the extent of geographic representation, and the years reported. Since FTR (1974) and Fay (1978) were often unable to determine how sample populations were selected, since they often could not identify the editions or forms of tests used or whether different forms were used by individual states from year to year, and since many states report on state-developed tests, the mixture of results reported in a diversity of statistical approaches add up to little that is certain—even if they did depict clear trends, which they do not. In general, the data within grades for specific states do not suggest clear trends, but rather tend to fluctuate from year to year.

The reported tendency of test scores to rise in the mid-1960s and then decline thereafter can be read in these data, in most cases, *only* by ignoring data for years that clearly contradict that trend and by noting very small declines after comparing the final year reported to the first year reported. Since much of the state data are for time spans after the mid-1960s, it is in-

TABLE 5.12

Data in Mean Grade Equivalents from Testing in School District E: Stanford Achievement Test, for years 1956-1962

Grade	Month/Season & Year Tested		# of S's	Reading Para. Mean	Word Mean	Language
3	Jan.	1956	8,782	3.65	3.49	–
	Jan.	1957	8,581	3.65	3.55	–
	Jan.	1958	9,248	3.93	3.78	–
	Jan.	1959	9,106	3.63	3.63	–
	Jan.	1960	10,065	3.91	3.99	–
	Jan.	1961	12,558	4.10	3.93	–
	May	1962	16,946	4.16	4.20	4.71
4	Sept.	1957	8,728	3.96	3.86	–
	Sept.	1958	8,871	3.95	4.12	–
	Sept.	1959	9,034	4.11	4.21	–
	Sept.	1960	10,064	4.27	4.21	–
	Sept.	1961	12,576	4.00	4.17	–
	May	1962	12,842	5.08	5.11	6.06
5	Sept.	1957	8,831	4.72	4.95	–
	Sept.	1958	8,451	4.90	5.08	–
	Sept.	1959	8,979	4.88	5.16	–
	Sept.	1960	9,001	5.07	5.17	–
	Sept.	1961	9,839	4.95	5.32	–
	May	1962	9,572	6.40	6.63	6.98
6	May	1956	6,606	7.24	7.32	7.26
	May	1957	6,659	7.09	7.55	7.42
	May	1958	6,972	7.72	7.80	7.48
	May	1959	8,379	7.81	7.93	7.61
	May	1960	8,204	8.08	7.59	8.05
	May	1961	8,816	7.72	7.69	7.62
	May	1962	8,968	7.87	7.65	7.47
8	Fall	1956	6,461	8.12	8.43	8.34
	Fall	1957	6,146	8.65	8.83	8.05
	Fall	1958	6,193	8.37	8.75	8.45
	Fall	1959	6,575	8.36	8.92	8.86
	Fall	1960	8,148	8.86	9.08	n.a.
	Fall	1961	8,177	8.22	8.76	8.51
	Fall	1962	8,943	8.18	8.83	8.94
	Fall	1963	9,011	8.21	8.86	8.62
	Fall	1964	9,818	8.54	8.95	8.94
	Fall	1965	11,676	8.40	8.90	8.60

Source: Farr, et al, 1974.

TABLE 5.13[b]
Data in Mean Grade Equivalents from Testing in School District E:
Iowa Tests of Basic Skills for years 1966-1972

Year	Vocabulary				Reading Comprehension				Composite Score[a]			
	Grade											
	3	4	5	6	3	4	5	6	3	4	5	6
1966	3.67	4.51	5.53	6.53	3.86	4.64	5.61	6.58	3.91	4.80	5.80	6.80
1967	3.72	4.50	5.55	6.45	3.84	4.57	5.58	6.40	3.89	4.78	5.77	6.62
1968	3.71	4.55	5.59	6.55	3.84	4.62	5.60	6.44	3.87	4.80	5.77	6.67
1969	3.65	4.50	5.46	6.38	3.78	4.61	5.53	6.50	3.85	4.76	5.70	6.65
1970	3.72	4.51	5.57	6.49	3.82	4.50	5.50	6.37	3.88	4.72	5.68	6.58
1971	3.45	4.25	5.20	6.13	3.53	4.31	5.21	6.20	3.54	4.42	5.36	6.30
1972	3.40	4.13	5.26	6.14	3.49	4.14	5.14	5.96	3.51	4.27	5.27	6.11

[a]Includes two arithmetic test scores.
[b]Numbers of students ranged from 15,000 to over 20,000.

Source: Farr, et al, 1974.

TABLE 5.14[b]
Data in Mean Grade Equivalents from Testing in School District F:
Iowa Tests of Basic Skills for years 1966-1970

Year	Vocabulary			Reading Comprehension			Composite Score[a]		
	Grade								
	4	6	8	4	6	8	4	6	8
1966	3.3	5.2	7.2	3.4	5.2	7.1	3.5	5.2	7.2
1967	3.1	5.2	7.0	3.2	5.1	6.8	3.4	5.1	7.0
1968	3.2	5.0	6.9	3.1	5.1	6.6	3.4	5.1	6.8
1969	3.2	4.9	6.8	3.2	5.0	6.6	3.4	5.0	6.7
1970	3.3	5.1	6.8	3.2	5.0	6.6	3.5	5.1	6.8

[a]Includes arithmetic subtest.
[b]Numbers of students ranged from over 10,000 to over 17,000.

Source: Farr, et al, 1974.

TABLE 5.15[a]
Data in "Mean Converted Score Units" from Testing in
School District F: Reading Subtest of Sequential Tests
of Educational Progress for years 1966-1970

Month/Year	Grade 10	Grade 12
October, 1966	279	292
October, 1967	279	291
June, 1968	277	290
June, 1969	275	289
June, 1970	274	288
National Mean	284	294

[a]Numbers of students ranged from over 6,000 to over 12,000.

Source: Farr, et al, 1974.

teresting that no clear declining trend can be established from the data, which are presented below by state alphabetically. It should be noted that although neither FTR nor the Fay study acquired comparative data on Indiana, the 1978 study (already discussed) provided a long-range trend comparison for grades 6 and 10 in Indiana.

Alabama. Alabama reported reading achievement scores for grades 8 and 11 from 1964-1971 to FTR. The data were in raw scores on the vocabulary and comprehension subtests of unspecified forms of the California Achievement Tests (CAT). Overall decline for comprehension for both grades in the 7 years was about one-half of one raw score point or less, with fluctuations over the period. Vocabulary scores rose very slightly or remained essentially the same.

Fay acquired Alabama data for 1971-1975 in grade equivalents for grades 4, 8, 10, and 11. The 1970 CAT was used. Data for some grades in certain years were unavailable. The only grade to record a change in grade equivalent was the tenth—a .2 decline between 1972-1973 and 1973-1974.

California. California's assessment program reported in May, 1979 that reading achievement scores for the state's second and third graders had risen slowly since 1970. Reported in mean percentiles for the state as compared to national percentiles on the Comprehensive Tests of Basic Skills (CTBS), third graders rose from the fiftieth percentile in 1970-1971 to the fifty-fourth percentile in 1971-1974, to the fifty-sixth percentile in 1974-1976, and to the fifty-seventh percentile in 1976-1978.

California sixth graders dropped on the CTBS from the forty-sixth to the forty-fourth pecentile between 1970–1971 and 1971–1972. They held that rank through 1973–1974, then rose to the forty-eighth percentile in 1974–1975, to the fifty-second in 1975–1976, to the fifty-third in 1976–1977, and to the fifty-fifth in 1977–1978.

In grade 12, percentiles on the reading section of the Survey of Basic Skills: Grade 12 dipped from the forty-ninth in 1970–1971 and 1971–1972, to the forty-seventh in 1972–1973 and 1973–1974, to the forty-third in 1974–1975 and 1975–1976, and to the forty-second in 1976–1977 and 1977–1978. A breakdown of the skills tested (vocabulary, literal comprehension, interpretive–critical comprehension, and study–locational skills) shows that the largest losses for the last year-span reported here were, interestingly, in vocabulary (4% of number of questions) and literal comprehension (4%) while the smallest loss was in interpretive–critical comprehension (1%); there was a 1% gain in study–locational skills.

Delaware. Delaware reported to Fay (1978) on an assessment program that began in 1972 for grades 1, 4, and 8. A modification of the Cooperative Primary Tests was used for grade 1 and the School and College Ability Tests were used for grades 4 and 8. The state summarizes trends in Delaware scores relative to national norms: continual increase for grade 1, with substantially higher performance since 1974; improvement for grade 4 since 1973; stable but low performance for grade 8.

Georgia. Fay (1978) acquired 1973–1976 data from Georgia on grades 4 (Level 10) and 8 (Level 14) on Form S of the Iowa Tests of Basic Skills (ITBS). A random sample of 15 students for each grade were drawn from each of 170 schools for a total of 2541 fourth graders and 2530 eighth graders. There is no appreciable change in reading test performance over the four-year period represented by these test scores (Table 5.16). It should

TABLE 5.16
Georgia Grades 4 and 8 Performances on Iowa Tests of
Basic Skills, Form S, in Grade Equivalents for 1973–1976

Grade	1973	1974	1975	1976
4	3.48/61.3	3.47/61.1	3.55/62.5	3.61/63.3
8	6.95/93.8	6.96/93.7	6.96/93.7	7.02/93.7

Source: Fay, 1978.

be noted that Georgia students characteristically perform below national
norms.

Hawaii. Hawaii reported to FTR on a statewide testing program using
three tests; however, only limited data from two of these were available.
Data for five grades are reported in midpoint percentile rankings for the
years 1965–1971 on the Sequential Tests of Educational Progress (STEP).
Generally the Hawaii scores are stable with a tendency to dip in 1966–1967
and 1970–1971 (see Figure 5.1). Interestingly, the score fluctuations are
fairly consistent across the grades, suggesting that some form of systematic
bias may have occurred.

On the upper primary—appropriate for grade 2—Form W (1963) of the
California Reading Test (CRT), second graders in Hawaii scored at the
sixty-ninth percentile from 1965–1966 through 1968–1969, but dropped in
1969–1970 to the sixty-sixth and in 1970–1971 to the sixty-second percentile.

Idaho. Eleventh grade data on four subtests of the Iowa Tests of Educational Development (ITED) were available as standard scores from Idaho
between 1960 and 1971. On three subtests, a very slight overall decline can
be seen after slight intervening increases (see Table 5.17). Overall,
vocabulary rises very slightly to end up at its 1960–1961 level. The average
reading scores were not available from 1960–1961 to 1965–1966. They
declined about one standard point between 1966–1967 and 1971–1972.

Iowa. Iowa has tested and recorded scores more intensely than any other
state in the nation. Iowa reported the results of a then-and-now study for
the years 1940 and 1965. It used the same tests both years on about 38,000
pupils in grades 3–8. These Iowa data are reported in median standard

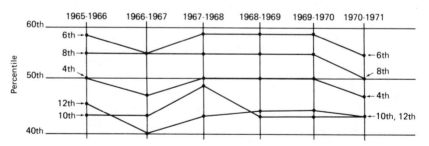

**FIGURE 5.1. Midpoint percentile rankings for Hawaii—fourth, sixth, eighth, tenth, and
twelfth graders on the *Sequential Tests of Educational Progress*, 1965–1971. Source: Farr
et al., 1974.**

TABLE 5.17
Idaho Average Standard Scores for Four Subtests of Iowa Tests of
Educational Development for Grade 11 in 1960-61 Through 1971-72

Year	Reading Social Studies	Reading Natural Science	Reading Literature	General Vocabulary	Average Reading Score	Number of Students Tested
1960-61	16.6	17.2	16.5	17.0	dna[a]	6,545
1961-62	16.8	17.5	16.6	17.1	dna	6,024
1962-63	17.0	17.7	17.0	17.5	dna	6,685
1963-64	17.3	17.8	17.0	17.7	dna	8,336
1964-65	16.9	17.5	16.5	17.4	dna	9,450
1965-66	16.8	18.2	16.8	18.0	dna	9,580
1966-67	16.8	18.2	16.9	17.9	17.3	10,994
1967-68	16.7	17.9	16.6	17.8	17.1	10,938
1968-69	16.7	18.0	16.6	17.9	17.1	10,252
1969-70	16.4	17.9	16.4	17.7	16.9	10,512
1970-71	16.4	17.4	15.8	17.4	16.5	9,147
1971-72	15.9	17.0	15.4	17.0	16.1	7,060

[a]Data not available.

Source: Farr, et al, 1974.

scores. There were gains in both reading and vocabulary for all grades in the 15-year span (see Table 5.18).

Another more recent report from Iowa[7] begins in 1955, when the first multi-level edition of the ITBS was used. The state education authorities report that Iowa norms are based upon a highly stable sample of par-

TABLE 5.18
Comparisons of Median Standard Scores on the Reading and the Vocabulary
Subtests of the Iowa Every Pupil Test of Basic Skills between 1940 and
1965 for the State of Iowa, Grades 3-8.

Grade	Reading			Vocabulary		
	1940	1965	Difference	1940	1965	Difference
3	37.6	43.7	6.1	38.3	41.6	3.3
4	46.8	55.4	8.6	45.6	50.7	5.1
5	56.8	64.5	7.7	56.3	60.4	4.1
6	64.4	76.2	11.8	64.3	76.7	12.4
7	76.0	85.3	9.3	76.1	84.0	7.9
8	86.0	93.7	7.7	84.1	91.0	6.9
Mean Differences			8.5			6.6

Source: Farr, et al, 1974.

[7] Iowa Department of Public Instruction, 1979.

ticipating schools from year to year. Over the past 10 years (1967–1977), between 93 and 97% of Iowa public school systems and approximately 95% of private and parochial schools have participated annually. These data were reported in average grade equivalent gains between the first year and the last year in 5-year periods and were scaled to 1965 as a base year. Between 1955 and 1960, achievement improved markedly. The Iowa Department of Public Instruction hypothesizes that this considerable gain is attributable to "extensive reorganization in the Iowa schools" during those years. Between 1960 and 1965, gains continued at a somewhat slower rate. Between 1965 and 1970, there was a general decline in all areas except work study. Losses in language usage were pronounced. Between 1970 and 1975, there were further losses, particularly in language skills. In 1976 and 1977, there were small gains in study skills. Changes in average grade equivalents (in months) between 1955 and 1977 are reported in Table 5.19.

These changes reflect a tendency for achievement test score gains to peak in the late to mid-1960s and then to decline until a slight recovery in the early to mid-1970s. In making this generalization, however, it must be noted that there are data reported in this chapter which do not reinforce such an interpretation.

Harnischfeger and Wiley (1975) give extensive breakdowns by grade and subtest for the state of Iowa ITBS between 1965 and 1975, calling the test scores "national data" and interpreting them as "consistent drops between 1963 and 1970 in a majority of subscales." They say that "The pattern is one of general increase, on all sub-scales from 1955 to 1963, consistent with the earlier finding of large gains to 1965. However, the national data show consistent drops between 1963 and 1970 in a majority of the subscales: Reading, Language, and Mathematics Skills [p. 50]." These national norming data are not cited or reported by Harnischfeger and Wiley (1975), however. Rather, data for the state of Iowa are presented.

The ITBS reading scores are given in total months for grade equivalents (see Table 5.20). Thus the gains and losses are in months.

These scores show considerable fluctuation within a fairly tight range over the years. In the 10-year span, there is a .7 gain for grade 3, a .4 loss for grade 4, a 1.9 loss for grade 5, a 2.5 loss for grade 6, a 5.2 loss for grade 7, and a 5.5 loss for grade 8. The data in Table 5.19 indicate that there were no losses between 1975 and 1977. The step line in Table 5.20 attempts to suggest when declines might be read as beginning for each grade.

One further then-and-now score comparison from Iowa[8] uses 1942 and 1977 average test scores on the Iowa Tests of Educational Development. Although there is no single reading subtest on the ITED, several are

[8] Data from Dr. William Coffman, Department of Public Instruction, State of Iowa (Release 4210-D25701-9/79).

TABLE 5.19
Changes in Average Grade Equivalents (in Months), Grades 3-8, 1955-77, In Iowa on the Iowa Tests of Basic Skills

Year	Test V Vocabulary	Test R Reading	Test L-Language Skills					Test W-Work Study Skills				Composite[a]
			Spelling	Capitalization	Punctuation	Usage	Total	Maps	Graphs	References	Total	
1955-60	3.5	3.9	4.2	8.3	3.3	5.6	5.4	4.2	4.6	4.1	4.3	4.2
1960-65	.9	.9	1.5	3.1	1.9	1.7	2.0	1.4	.7	1.5	1.2	1.3
1965-70	-.7	-1.1	-.7	-1.6	-1.3	-2.2	-1.4	.4	.3	.4	.4	-1.0
1970-75	-1.0	-1.4	-1.7	-3.6	-3.0	-2.5	-2.7	-1.3	-.6	-1.2	-1.0	-1.5
1975-77	.1	0.0	.1	.7	.6	-.2	.3	.3	.5	.2	.3	.1

[a] Includes mathematics subtests scores.

Source: Iowa State Department of Public Instruction, 1979.
Supplied by Iowa Testing Programs, University of Iowa

115

Roger Farr and Leo Fay

TABLE 5.20
Iowa Tests of Basic Skills Reading Subscale data in "Base-Year
Grade Equivalents" for 1965-1975 for Iowa[a]

Grade	3	4	5	6	7	8
1965	34.5	44.5	54.5	64.5	74.5	84.5
1966	34.8	44.9	54.7	64.5	74.2	84.6
1967	35.2	44.8	54.8	64.3	73.8	83.8
1968	35.2	45.0	54.6	64.7	72.8	83.4
1969	34.6	45.0	54.5	64.2	72.6	82.8
1970	34.7	44.3	54.4	63.6	71.4	82.0
1971	34.8	44.2	54.0	64.0	71.4	81.6
1972	35.0	44.4	53.9	63.7	71.9	81.8
1975	35.2	44.1	52.6	62.0	69.3	79.0

[a] The figures are for total months; to get the grade
equivalent, one divides by 10, as they are figured on
a 10-month school year.

Source: Revised from Figure 24, p. 53, in Harnischfeger
and Wiley, 1975.

language-related and it is assumed that in most, reading ability is required.
These data demonstrate no decline trend. Norm scores are lumped for
grades 10 and 11 (see Table 5.21). The report concludes that (*a*) even
though the proportion of the high school group in school has increased,
there has been no drop in overall achievement; (*b*) where there has been a
decline in subscore, it has been very small; (*c*) where there has been an increase, it has been relatively large; and (*d*) the areas of increase suggest
reasonable differences in emphasis.

Michigan. Michigan has tested all fourth and seventh graders since
1972–1973 with state-developed tests. The fourth-grade test measures 19
performance objectives, the seventh grade measures 20, with five questions
each. The Michigan data were reported to Fay (1978) in percentages of
pupils attaining proportions of objectives (see Table 5.22). The shift from
the lower three quartiles to the 75–100% attainment is quite marked for
grade 4 and evident for grade 7. Researchers should note that the Michigan
program has also gathered accompanying demographic and other societal
data with the test data.

TABLE 5.21
Changes in Average Test Scores on the Iowa Tests of Educational
Development for the Years 1942 and 1977 in Iowa

Subtest	1942 Norm Grades 10 and 11	Average of 10 and 11, 1977
Expression	15.0	14.95
Quantitative Thinking	15.0	14.70
Social Studies	15.0	14.75
Natural Science	15.0	16.05
Literary Material	15.0	14.70
Vocabulary	15.0	15.75
Sources of Information	15.0	16.00
Composite	15.0	15.80

Source: Dr. William Coffman, Department of Public Instruction,
State of Iowa (Release 4210-D25701-9/79).
Supplied by Iowa Testing Programs, University of Iowa

Minnesota. Harnischfeger and Wiley (1975) report scores from the administration of the Minnesota Scholastic Aptitude Test (MSAT) between 1958 and 1972 to Minnesota grade 11 students. Total mean scores on the test begin at about 29.5 in 1958, rise sharply to about 35 by 1966, and then decline to about 32 by 1972. Thus, in the 15-year period, there is an overall gain of about 2.5 points, but between 1966 and 1972, there is a loss of 3 points. This tends to reflect what has been called a general peaking period for test scores in the mid-1960s; but it could also raise the possibility of considering and studying the mid-1960s as an atypical period of gain.

New Hampshire. New Hampshire reported using three tests to FTR (1974) over different time periods, but only eighth-grade data from 1959 and 1962–1964 on two Metropolitan Achievement Test (MAT) subtests were reported (see Table 5.23). New Hampshire students who scored in specific percentiles on the test had higher grade equivalents than the students who scored in equivalent percentiles nationally. Trendwise, the grade equivalents for New Hampshire students increased in word knowledge until 1963 with a drop in 1964 and in reading through 1964 except in the tenth percentile.

New Jersey. New Jersey reported its 1976 rating to Fay (1978) on a total of 10 reading subscores in a state-developed test, administered since 1972. For

TABLE 5.22
Percentages of Michigan 4th and 7th Graders Attaining Proportions of
Objectives on State-Developed Reading Test

Year	Grade 4 Proportion of Objectives Attained				Grade 7 Proportions of Objectives Attained			
	.00-.24	.25-.49	.50-.74	.75-100	.00-.24	.25-.49	.50-.74	.75-100
1973	29.3	13.7	18.6	38.5	21.9	13.2	14.9	50.0
1974	21.7	12.2	17.5	48.6	20.2	12.0	12.5	55.2
1975	20.9	11.3	16.4	51.4	20.4	11.1	12.0	56.6
1976	18.7	10.6	15.3	55.4	20.2	11.9	12.3	55.6

Source: Fay, 1978 from Michigan Department of Education. Status Report
on Michigan Basic Skills Education. (n.d.).

the fourth grade, the state judged improvement-performance as either
"satisfactory" or "very satisfactory" on seven subskills. At the seventh
grade, five were rated "satisfactory" (a score scale of 71-85), one "very
satisfactory" (a score scale of 86-100), and one "inconsistent." At the
tenth grade, only three subskills were rated "satisfactory."

In a special report (Golub-Smith, 1979), New Jersey reports comparison
mean raw reading scores for the three grade levels on the state test, explain-

TABLE 5.23
New Hampshire Grade Equivalent Scores on the Metropolitan Achievement Test,
Advanced Battery: Forms A, B, and C. (Scores reported for subtests of Word
Knowledge and Reading only for eighth graders)

Percentiles		1959 Grade Equivalent	1962 Grade Equivalent	1963 Grade Equivalent	1964 Grade Equivalent	National Norm
Word Knowl- edge	90th	10.0+	10.0+	10.0+	10.0+	10.0+
	75	10.0+	10.0+	10.0+	10.0+	10.0+
	50	9.1	9.5	9.9	9.5	8.1
	25	6.7	6.7	7.8	7.0	6.0
	10	5.4	5.4	6.0	5.5	5.1
Read- ing	90	10.0+	10.0+	10.0+	10.0+	10.0+
	75	10.0+	10.+	10.0+	10.0+	9.9
	50	8.0	8.5	8.5	9.2	8.0
	25	6.0	6.6	6.8	6.8	6.0
	10	4.7	5.3	5.1	5.1	4.7
Number Tested			2,632	2,243	4,287	

Source: Farr, et al, 1974.

ing that after 1976, random samples were substituted for testing of all students.

The data for 1976 vary slightly between the two sources. Substituting the Golub-Smith 1976 data for that reported to Fay, the two reports are combined in Table 5.24, which demonstrates the difficulty in reading overall trends into such data. For grade 4, however, the gain trend is obvious.

New York. In the FTR study, New York reported scores for grades 3 and 6 on state-designed tests covering the years 1966–1971 (see Table 5.25). Over the 6-year span, there was a net loss of about a half a raw score point for grade 3 and 2 raw score points for grade 6. These declines are slight when one notes that the standard deviations increased for the period, indicating a more diverse population.

Fay's study (1978) includes data showing percentages of third- and sixth-

TABLE 5.24
Reading Mean Score Data from New Jersey (1972, 1974–1978) for Grades 4, 7, and 10 of the State's Educational Assessment Program

Year/Form[a]	No. of Cases	No. of Items	Mean Score	Standard Deviation
		Grade 4		
1972	111,177	80	57.24	15.56
1974	105,849	96	74.77	17.47
1975	99,788	95	79.22	14.13
1976	95,258	95	81.65	12.97
1977	327[b]	95	83.14	11.72
1978	383	95	82.90	12.69
		Grade 7		
1974	110,782	86	65.29	15.98
1975	109,914	90	67.20	17.27
1976	108,864	90	67.60	16.55
1977	345[b]	90	66.80	16.98
1978	321	90	73.36	12.31
		Grade 10		
1974	107,577	80	58.83	14.51
1975	108,207	80	61.11	13.95
1976	109,427	80	61.41	14.49
1977	335[b]	80	62,32	13.38
1978	331	80	59.63	15.71

[a]New Jersey modified the test yearly up through 1976; the same test was used in 1976–1978.

[b]In 1977, a random sampling testing began.

Source: Golub-Smith, 1979, derived from several figures.

TABLE 5.25
Mean Raw Score for Third and Sixth Graders in New York on the New York
State Elementary School Reading Test and the New York Minimum
Competency Reading Test for the years 1966-1971

Year	Grade 3			Grade 6		
	Mean Raw Score	Standard Deviation	Mean Percentile Rank	Mean Raw Score	Standard Deviation	Mean Percentile Rank
1966	31.90	12.27	48	41.97	13.90	45
1967	31.82	12.50	48	41.71	14.02	45
1968	32.14	12.40	48	41.80	13.96	45
1969	31.78	12.57	48	40.86	14.20	43
1970	31.51	12.34	47	40.18	14.27	41
1971	31.38	12.51	47	39.94	14.30	41

Source: Farr, et al, 1974 (Revised).

grade students in New York state who scored above a 23% reference point
on a combination of unnamed standardized tests. The data are for the
years 1966, and 1971-1975. For third grade, the percentage rises from 77 in
1966 to 81 in 1975. For sixth grade, it drops from 77 in 1966 to 73 in 1971
and then only one point, to 72, by 1975. Reading even a slight trend into
these data is risky, and assigning causes without accompanying educational
and societal data would be irresponsible.

Ohio. Ohio reported to FTR for the years 1965-1970 on a testing program
but did not name the tests used or report how the samples were drawn.
Mean raw scores are given for an average of over 50,000 pupils in each of
grades 4, 6, 8, and 10. The mean score for grade 4 peaked in 1967, and
1970's score was 3 points higher than 1965's. For grades 6, 8, and 10, there
were 5.5, 3.2, and 5.4 point total declines respectively for the five years.
The only declines that suggest a trend are for grades 8 and 10 and average
about one-half a raw score point per year.

South Carolina. Although South Carolina did not provide Fay with data,
the state reported that between 1970 and 1975, the percentage of student
scores in the lowest quartile on the Comprehensive Tests of Basic Skills
(CTBS) had declined steadily, but that "8 to 21% more than would nor-
mally be expected scored below the national fiftieth percentile" in the areas
of reading and language arts.

Virginia. Using the SRA Achievement Tests for grades 4, 6, and 8, and the
Sequential Tests of Educational Progress (STEP) for grades 9 and 11,

Virginia reported national percentile equivalents within the years 1970–1975. Percentiles for grades 4, 9, and 11 fluctuated around 50 although the SRA test was used in 1973–1974 and 1974–1975 at grade 11. Only 1972–1973 through 1974–1975 were reported for grade 6 (a 6 point percentile gain) and 1973–1974 and 1974–1975 for grade 8 (a 2 point percentile gain). No trends are obvious from these data (reported to Fay).

West Virginia. West Virginia reported to FTR on Stanford Achievement Test scores for grade 3 from 1964–1969 and for grade 6 for seven periods between 1957 and 1969. No pattern of decline or gain can be read into these data, which are quite stable. Mean standard scores on the STEP for grades 9 and 11 were also reported between 1963 and 1968 for an average of 30,000 students in each grade. The scores fluctuate with only about a 1-point decline in each grade (see Table 5.26). Two earlier periods are reported for grade 12. Again, any trends are very slight, and there is a general tendency for scores to peak at about 1966–1967 before dropping off.

Summary of State Data. The data reported in this section should serve, above all, to stress what a mixed bag available data make. The number of states covered, the short and uneven time spans represented, the variety of tests used, and the mix of declines and gains themselves suggest that drawing any sure trends from the data is certainly precarious if not irresponsi-

TABLE 5.26
Mean Grade Equivalent Scores for 3rd and 6th Graders on the Reading Subtest of the Stanford Achievement Test in West Virginia for the Years Spanning 1957 through 1968–69[a]

Year	Grade 3		Grade 6	
	Word Meaning	Paragraph Meaning	Word Meaning	Paragraph Meaning
1957			5.8	5.1
1962–63[b]			6.2	6.2
1964–65[b]	3.15	3.24	6.09	6.12
1965–66[b]	3.00	3.08	5.3	5.6
1966–67	3.28	3.35	5.76	6.12
1967–68	3.30	3.39	5.77	6.14
1968–69	3.32	3.39	5.71	6.05

[a]Over 2,000 sixth graders were tested in 1957; all other data is based on samples of over 30,000 subjects.
[b]Reported in state medians, not means.

Source: Farr, et al, 1974.

ble. In addition, the reports are often void of indications of changes in tests or forms used and are sometimes void of test identification. Sample selections are often unexplained; any attempt to interpret scores along with factors like age changes, dropout rates, demographic shifts, etc. is rare; statistical analyses are often questionable and sometimes nonexistent; and reporting is often limited.

Other Nationally Administered Tests

There has been a general contention that a decline in test scores (on all measures) nationally administered can be documented. Harnischfeger and Wiley (1975) and FTR (1974) are frequently cited as dependable meta-analyses of the data available up to the mid-1970s. Yet a close analysis of the metropolitan and state data in FTR underlines their warning that weak data allow only weak inferences.

The bulk of the data supporting declines in reading performance among U.S. students comes from scores on college entrance exams. While the decline on such aptitude tests as the Scholastic Achievement Test (SAT) cannot be discounted, the sweeping conclusions drawn from these score declines about reading achievement at all grades are not scientific. As an indicator of performance at grades 11 and 13, the decline may be less pronounced than generally believed, and there are other related test data for these grades that do not reinforce the alarm about the decline.

The SAT Decline

Declines in standard scores of the SAT and *American College Test* (ACT) have been discussed frequently, as well as elsewhere in this book. Generally, on the verbal section of the SAT there has been a decline of 45 standard points between 1960–1961 and 1976–1977 for high school juniors and seniors (Wirtz, 1977, p. 6).

It should be noted in any discussion of the SAT decline that standard score points potentially exaggerate the decline. A test of approximately 80 raw score points, the SAT verbal section is reported in a standard score scale ranging from 200 to 800. The ratio of standard score points to raw score points varies from year to year and also varies at different points on the score distribution scale. An average of approximately 10 standard score points to each raw score point is typical of the middle range of test scores. Thus, a 45 standard score point decline is equivalent to about 4 raw points, or less than a third of a raw score point a year.

Moreover, the report of the Advisory Panel on the Scholastic Aptitude Test Score Decline (Wirtz *et al.,* 1977—hereafter called the Wirtz Commission report) credits three-fourths of the decline up to 1970 to the change in the group of students taking the test, and one-fourth of the decline after 1970 to that factor. Based, then, on the declines between 1961 and 1970 and between 1970 and 1977, over 1.5 of a 4.5 raw score decline is attributed to sample change, and 3 raw points (.2 per year) to actual decline. Stated simplistically, as college has become a more real opportunity for many more students, the SAT has been taken by a sample which has included more students of lower abilities.

This analysis makes no attempt to weigh any of the other societal factors considered by the Wirtz Commission, whose cautiously responsible report and very interesting Appendixes go relatively uncited (and possibly unread) by the discussants of the SAT decline issue.

A point related to the consideration of the decline in college entrance aptitude exam scores seems pertinent: The verbal section of such tests measures sophisticated language aptitudes. Upon examining the vocabulary tested by the SAT, for example, one has to acknowledge that it constitutes an erudite lexicon. The multiple-choice comprehension questions engage the testee in sophisticated thinking exercises that require a keen understanding of syntax, an experienced knowledge of human nature, an awareness of prevailing value systems, and an ability to deduce by using these skills to eliminate the obliquely correct options and to select the somewhat more precise correct option (Farr & Olshavsky, 1980).

This point is made because these tests, which are designed to assess aptitude for success in higher learning, have been nationally identified with some unexplicated definition of "basic literacy." This has resulted in the 3-point raw score decline being used as an unchallenged rationale for educational change that is designed to hopefully improve some sort of basic literacy training. To date, the contents of the many state tests and school programs resulting from such response to the SAT score decline have not been content-analyzed to see how, or if, the tests relate to the kind of thinking measured by college entrance exams.[9]

The response to the SAT score decline appears to have completely ignored the warnings of the Wirtz Commission and of the test's publisher. The Commission prefaced its report with:

1. "Any generalization from the SAT statistics has to be carefully qualified. It should not be extended to cover the situation of American youth as a whole or the overall effectiveness of the learning process."

2. The guidelines to the use of the SAT "warn sharply against misuse as

[9] This is supported by findings cited by Jaeger (Chapter 10, this volume) that suggest that just the opposite outcomes may be emerging (Eds.).

measures of the broader effectiveness of elementary and secondary education in general [p. 5]."

The reason for these warnings is that a test designed to measure the aptitude of college-bound juniors and seniors is deliberately concerned with higher levels of comprehension. Thus, the SAT decline raises a more reasonable concern about the success of our schools in teaching higher levels of comprehension. This point tends to be supported by the Indiana study (Farr, Fay, & Negley, 1978). In that study gains on comprehension subtests were smaller as the tasks became more global (see Tables 5.3 and 5.4)—word comprehension gains were greater than those in sentence or paragraph comprehension and sentence comprehension gains were larger than those in paragraph comprehension. Assuming that comprehension of broader segments of text and inferential comprehension are related as higher levels of comprehension, the NAEP data also tend to support this concern. NAEP gains on literal comprehension for all three age groups were offset somewhat by slight losses in inferential comprehension for the 13- and 17-year-olds.

In considering this conclusion, however, it is interesting to note a comparison study by Torsten Husen (Wolf, 1977). It showed that the top 9% of seniors in U.S. schools performed better in reading comprehension and in understanding and interpreting literature than did similarly elite groups in any of 22 other countries studied. These countries included England, Sweden, Japan, West Germany, and Belgium.

The issue is further clouded by the fact that samples of high school juniors who took the Preliminary Scholastic Aptitude Test (PSAT) showed no score decline of the size on the SAT between 1963 and 1973. After 1973, declines on the two tests were parallel. On a battery of subject area achievement tests, which juniors and seniors can elect to take in conjunction with the SAT, there have been, in the last ten years, only small declines in four areas and increases in six. These tests tend to be taken by high achievers, however, and none is a direct measure of reading (Wirtz, 1977, p. 22).

While there have been declines parallel to the SAT on the Graduate Record Exam (GRE), which is taken by college graduates seeking entrance into programs that lead to post-graduate degrees, it is interesting to note that there have been score increases on entrance exams administered to prospective law and medical students. It should be noted, too, that the GRE is scaled to 2095 students who were tested in 1952 and who were "not representative of any population; that percentages for the GRE are rescaled every 3 years, limiting longitudinal comparisons; and that the content of the GRE was changed substantively in the early 1970s [Tyler, 1972, pp. 667–668]."

CAVEATS ACCOMPANYING TEST DATA

None of the data presented so far should be considered without some understanding of the serious caveats accompanying them. Determining reading achievement trends for the nation over time is difficult. It is difficult even if the geographic focus is more local than national, for key factors in the comparison shift or change, and thus, the factors for one (earlier) time period are not the same as for another (later) time period.

Subject Matching

Finding ways to adjust for differences in student samples between time periods is a difficult if not impossible task. Highly scientific sample selection, combining randomization with cautious stratification, can result in some valid comparisons between relatively short time periods. Thus, the National Assessment of Educational Progress (NAEP) study, which controls for key factors in selecting its national samples, is a relatively trustworthy comparison.

But in comparing samples from more divided time spans, a host of factors with potential impact must be considered. One can, for example, compare all the students in a certain grade level for a given metropolitan area, but over time, the make-up of the general population in that area may have (and usually has) changed considerably. Any number of economic and societal factors will have affected the student population and will have made the attitudes, abilities, and backgrounds of students in the later time period quite different from those of the students being studied years earlier.

A comparison of reading achievement levels on a given test for all third, sixth, ninth, and twelfth graders in Detroit's schools might be possible for the year 1950 and for the present, but the differences noted are more likely to describe economic and societal changes within the city than they are to describe how effectively reading is being taught and learned there.

That is not to argue that many educational factors, including reading instruction, are unlikely to affect differences in scores on the same test over time. Rather, the point is that educational differences are but a part of a very complex set of factors and that reading instruction is but one of the educational factors affecting performance on reading achievement tests.

The Indiana study (Farr, Fay, & Negley, 1978) helps illustrate the point. The performance of sixth and tenth graders in Indiana from 1944–1945 was compared to that of those in 1976 on the identical test. It was assumed that the populations from the two periods were a reasonable match because (*a*) the geographic area was state-wide, incorporating metropolitan, suburban, small town, and rural students; (*b*) the 1944–1945 testing had randomly tested nearly one-fourth of the students in each grade and could demonstrate an evenly spread geographic and demographic representation; and (*c*) the 1976 testing sampled thousands of students at each grade level using a stratified–random design to match the earlier sample.

Yet this study quickly uncovered a striking age difference between the students for the two periods at both grade levels. The students from the mid-1940s were months older than those in 1976. One factor that accounted for this difference was a change in retention policy.

Retention is a factor that has the largest potential impact on test performance, as it means more school as well as life experience for the students involved. The question is, how much of the age difference noted (10 months for sixth graders and 14 months for tenth graders) was due to retention of the 1944–1945 students? No study existed to reveal this, but a safe assumption is that much of it was.

Another educational factor was discovered in the Indiana study that could have had significant impact on the test performance of the tenth graders. Dropout decline helped raise the average number of years of school for an Indiana citizen from 7.5 in 1940 to 12.1 in 1976. Since the 1944–1945 student reached the legal drop-out age no later than between ninth and tenth grade, it is safe to assume that the tenth-grade class in 1976 included a significant percentage of students who would not have been in school in the 1944–1945 sample. Those students, we must assume, would not be among the strongest performers on an achievement test, and, therefore, they would lower the performance average of the whole sample group.

Dropout in the earlier period was influenced by societal and economic factors that illustrate their potential effect on achievement testing as well as on the complex relationships between educational and other factors. In 1944, World War II had put an end to unemployment, defense production was at its peak, and it was patriotic as well as immediately lucrative to consider dropping out of school to take a job in industry.

An on-going study (Blomenberg, 1980) of such factors for the two periods studied in Indiana is presently detailing factors that affected such changes as population increase and shift to urban areas as well as economic trends and technological development, such as those in transportation and communication. Blomenberg considers the impact of such factors on diverse state-wide educational changes, such as school consolidation, teacher certification, and curriculum development (see Table 5.27). Blomenberg summarizes that "as America was transformed, so too was Indiana. Juxtaposing the substance of selected social and economic variables during the 1940s and the 1970s substantiates this. Basic differences can be noted in [each of these variables]."

TABLE 5.27
Composite Picture, Socioeconomic Variables in Indiana, 1940 and 1970

	1940	1970
Demographic Information		
Total population	3,427,796	5,194,000
Percent urban	55.1%	64.9%
Percent white	96.4%	92.8&
Percent black	3.6%	6.9%
Percent Spanish heritage	--	0.3%
Labor Force Information		
Total labor force	1,331,378	2,129,859
Unemployment rate	13.5%	4.8%
Percent labor force female	20%	36.5%
Percent working women married	37%	60%
Occupation of Labor Force		
Percent professional/ technical	6.8%	11.8%
Percent farmers and agricultural workers	27.5%	2.9%
Percent clerical, sales, service workers	22%	32.2%
Education		
Median years schooling	7.5 years	12.1 years
Education past high school	9.0%	16.9%
Income		
Per capita income (1972 dollars)	$2300 (1944-45)	$4692.75 (1976)
Age		
Median age	30.3 years	27.2 years
Housing Characteristics		
Plumbing equipment	Outdoor toilets (45.6%) Indoor bathrooms (42%) No running water (38.4%)	Lacking some/all(3.3%) More than 1 bath (23%)
Kitchen equipment	Refrigerators (41.1%) Iceboxes (28.3%) Gas stove (42.8%)	Complete kitchens (96.4%) Food freezers (33.1%) Dishwashers (13.1%)
Lighting equipment	Electric (84%) Kerosene (15.3%)	Electric (100%)
Other home equipment	Wringer washer/tub[a] Radio (88.2%)	Automatic washer (56.6%) Automatic dryer (51.8%) Television (96.2%)
Automobiles	1 per household[a]	1 per household (51%) 2 per household (35%)

[a]Not counted in census.

Source: Blomenberg, 1980.

Without attempting to use them as score adjustments, the Wirtz Commission took careful note of such factors as the politically tumultuous impact that the 1960s had on youth (including their changing attitudes toward tests and higher education) and the influence of television. Other very relevant factors include teacher training, experience, and professionalism.

The major point here goes beyond the importance of drawing comparable population samples from different periods. Once the caveat is accepted, it should promote caution in interpreting achievement comparison of samples from different time periods. The factors that make comparable sampling extremely difficult also point up that how or how well reading is taught is but one in a complex network of potential explanations of performance differences.

Test Matching

One of the prime decisions a researcher who records achievement trends data must make is whether the identical test used to test an earlier population will be used to test later subjects. Frequently, the exact test used on earlier populations is unavailable, but when it is or can be duplicated, most researchers will opt to use the identical test and to note the potential disadvantage it creates for the contemporary student.

A test written for use in previous decades will likely contain content and emphases that were appropriate to the curricula of the time it was written. Likewise, it will not cover or reflect much of the emphasis that has been incorporated in the new curricula that attempt to recognize changing societal needs. The earlier test will not deal with some of the vocabulary or concepts that may play a major role in the life of a contemporary citizen, and it will not reflect connotative changes in words that have developed after it was written. Often, dated tests will reflect value systems unique to the period for which they were written, and to get some answers correct, the test-taker must assume the values of the earlier testwriters.

Some critics of education would readily admit that the mastery of particular concepts, information, and even values are among the key focuses they wish to see compared, but anyone who would scientifically compare achievement should acknowledge that as tests measure mastery of the skill and content developed by curricula, contemporary students who take dated tests are at a considerable disadvantage—just as the earlier students might be if it were possible to test them with a contemporary instrument.

Testmakers do not propose that their instruments are developed to promote curriculum change, but, rather, to measure mastery of curricula that they assume have been dictated by a developing society. Many tests that produce the data used in debates about our schools do not even do that, but instead they purport to measure the test-taker's *aptitude* to succeed in an expected curriculum.

It is appropriate that achievement data comparisons are used to show whether today's schools are developing the knowledge and skills that will best enable students to succeed in today's society. It is essential to this concern to recognize that yesterday's tests may include items with questionable validity for today's citizen and will not cover many emphases in curricula that developed after the test was written. The understanding of that possibility suggests the desirability of reorienting achievement trend discussions so that they focus on (*a*) how well students in each period performed on content and skills stressed in the school curriculum they experienced; and (*b*) how valid the curriculum of each time period was in terms of the functional demands made by society during the time the curriculum was in force. With this approach, score comparisons across time would have less significance, and it would make sense to use different tests.

The use of data from different tests to determine trends is equally problematic when one compares the performances of the same student at different levels to see if he or she is progressing as one would wish. This practice focuses the problem of mixing tests which invariably emphasize different skills, publisher to publisher, and different skills at different levels, even within test series from a single publisher.

A recent analysis of tests by six publishers of reading tests (Farr & Olshavsky, 1980) shows that there are considerable differences in the emphases of subskills measured by different tests and in the time allotted to those subskills. Comprehension was the only reading subskill measured by all six, and at grade 2, it comprised from 28 to 60% of the total score and 28 to 75% of the time allotted, depending on the publisher. The emphases of subksills within and across publishers varied markedly across test grade levels. Thus, comparisons using tests with such differences are about as scientific as comparing apples and oranges.

There is a trade-off in the decision of whether to administer earlier tests to contemporary students in evaluating achievement trend data. Most researchers decide that using the same test affords more reliable comparison even though doing so may create a disadvantage for the students in the later time period. Although it may need to rely on subjective judgment, it seems a good idea to eliminate this disadvantage by deleting items from earlier tests for the later administration when such items have become factually incorrect or when they stress content or values uniquely identifiable with the earlier period.

Renorming Data

Publishers who go to the greatest care and largest expense to norm and renorm editions of their tests are the first to suggest the limitations of norming–renorming data as trend data. They underline the following caveats that limit interpretations of the data:

1. The content of the two tests used is different; that is the basic reason for the renorming process. Test publishers rely on many facets of the education scene to determine what the tests should measure. They consult educators at all levels and analyze the content of texts. This approach attempts to identify curriculum emphases. Some content and skills taught at one grade level at an earlier time period may be identified as currently emphasized at earlier or later grades. Since the instructional objectives identified in this process change with time, it is desirable that the tests change to reflect and measure them, and comparing the results yielded by different measuring instruments is extraordinarily difficult.

2. The characteristics of the populations used between norming and renorming are unavoidably different, grade-for-grade. They are especially affected by differences in school entrance ages, retention rates, and dropout rates. Age differences have a high potential impact on test results. At upper levels, dropout rates can be very significant in explaining performance differences. Despite the fact that test publishers take great pains to obtain good normative groups, they cannot guarantee, or demonstrate, that the norms for various editions of a test are based on adequately comparable samples. Therefore, the testmakers are the first to caution that the grade equivalents of various editions cannot be assumed to have identical meaning.

3. Psychometric procedures change between norming and renorming (e.g., besides the problem of equating study designs and data, there may be differences in the construction of equating tables or there may be changes in defining or determining grade equivalents).

Renorming data, which are used frequently by critics of education to prove decline, could be cited here, but the caveats inherent in renorming data recommend looking elsewhere in an attempt to find reading performance trends.

Grade Equivalents

One of the greatest misfortunes in the attempt to determine reading trends from existing data is the fact that so many of the data are reported in grade equivalents (GE). As the discussion on norming data notes, comparing grade equivalents over time is highly risky because it attempts to shift objective and content emphases within tests to match the prevailing emphases in the schools. That is, if skill x is now emphasized on the test in grade 8 as opposed to grade 9 (where it was previously emphasized), that shift can have marked impact on the GE score of test takers and will depend on how well the test change really reflects the curriculum change. And if 1980 eighth graders score a GE of 8.6 on the current edition, it means little compared to an 8.4 or 8.9 scored by eighth graders previously.

Comparing GEs across grades is also misleading. A test that has no content more difficult than grade 6 level can yet produce a GE of 12.0 or more. That certainly doesn't mean a child who scores 12.0 can handle grade 12 material. Thus a GE is not indicative of an instructional level. And to compare one grade's mean GE on one level of a test to another grade's mean GE on a different level of a test is meaningless. The levels are usually derived from different tests, and there is no safe way to compare the GE produced on one to that produced on another. Yet the nature of the GE encourages just that kind of test score misinterpretation, in the classroom as well as in trend discussions.

LITERACY DATA TRENDS

Examining achievement test data is not the only valid approach to determining reading trends in the United States. Another type of information that might serve the effort is literacy data. The longest-range national literacy data have been collected by the U.S. Census Bureau, but there are many weaknesses in this information. With no acceptable definition of *literacy*, the means of judging literacy has shifted from census to census. The varying ways of determining literacy demonstrate the difficulty of drawing comparisons over time with these data (see Cook, 1977).

In 1910, a simple "yes" or "no" response to a literacy question in English determined if a citizen was literate. By the 1970 census, literacy was determined by whether one could read or write a simple message in any language.

This problem with defining literacy persists and has led to unsuccessful universal attempts to define it in more explicit terms, such as *basic literacy* or *functional literacy*. Nonetheless, the census data indicate a trend away from illiteracy, that is, when illiteracy is defined by the total inability to read or write. The census data show that in 1910, 10.7% of the U.S. population 10 years of age or older (6.2 million persons) were illiterate. By 1930, the number of persons over 10 years of age who could not read or write in any language was down to 4.3% (4.3 million). By 1959, a special census study showed that only 2.2% of the persons over the age of 14 could not read or write in any language. By 1970, that figure was down to 1%.

Other approaches to defining literacy (or functional literacy) have led to some startling figures that have been injected into the national discussion of whether our educational system is doing its job. Many of these data are irrelevant to determining trends, however, for no matter how literacy is defined, the figures derived from those definitions are for one time period only and thus have no longitudinal value.

The attention such studies have received, however, requires summarizing their findings and dealing with a recent critical and sophisticated reinterpretation of them.

Two studies (Harris, 1970, 1971) identified as illiterate 2 and 4%, respectively, of U.S. citizens who had completed 12 years of schooling and 11 and 52%, respectively, of the population who had completed 5 years of schooling. The first study used five general application forms requiring up to 15 responses each. The second study used 57 items drawn from actual application forms, advertisements, and telephone operation instructions. These tests were administered to a representative sample of the noninstitutionalized population over 16 years of age.

The Educational Testing Service ([ETS] Murphy, 1973) studied a representative sample of noninstitutionalized adults, 16 years of age or older, with 170 items. The respondent circled answers to samples of "typical adult reading." ETS found 19% of high school graduates and 14% of professional/managerial persons to be "illiterate."

There have been several other such studies, the most notable the Adult Performance Level Project ([APL] Northcutt et al., 1975). It selected items from five general areas of knowledge and classified each as one of 5 skills (3 were language-related). A representative sample of 18-year-olds living in "households" was tested. The research found that 11% of those who completed high school were illiterate, and that 76% of those who had finished grade 5 were "incompetent"—the study's equivalent to *functionally* illiterate. It found 11% of subjects in professional/managerial positions functionally incompetent!

A recent study (Hunter & Harman, 1979) reviews the problem of defining functional literacy. Equating the lack of a high school diploma to "disadvantaged," the study reports that "It is not surprising, therefore, to discover that most American studies of educational attainment use graduation from twelfth grade as the critical point [p. 27]." Noting the decline in school dropout, Hunter and Harman then report that if "functional literacy" is determined by whether one has a high school diploma, then 38% of the population 16 years of age and over would be labeled functionally illiterate. This study, however, is careful to note that high school completion is not a reliable indicator of functional literacy and that consideration of social, economic, geographic, racial, and ethnic factors are equally important. A thorough treatment of the "'soft" data on literacy leads one to the conclusion that the 38% figure is at least a target population, among which the nation's illiterates—whatever they number—may be found.

What may be more relevant to literacy trends is the Hunter and Harman data showing a dramatic rise in those graduating from high school between 1910 and 1963—from around 5 to about 75% of the population. The median level of school completion for the population 25 years of age and over has risen since; while it was between grades 9 and 10 in 1950, it has now surpassed the grade 12 level.

Another analysis (Fisher, 1978) presents some interesting, logical criticisms of the Harris and APL studies. It attacks the label *functional literacy*, arguing that those subjects tested who were in professional/managerial positions were apparently *functioning,* no matter on what level the test put them [p. 5]. Fisher argues that this "mislabeling" applies equally to high school graduates. Coupling this logic to adjustments for errors due to fatigue and test construction, Fisher reestimates the illiteracy rate among high school graduates to be 1–2% [p. 7]. From within the one-set limitations of the data for two studies (Harris, 1971; Murphy, 1973, 1975), Fisher draws trend data of a sort. Using ages 16, 16–24, and 25–29 as Harris age groups and 16–20, 21–29, and 30–59 for Murphy's, he shows score increases with age—an increase which he attributes to education.

Overall, literacy data suggest the following:

1. The United States can claim substantial progress in encouraging education and in reducing illiteracy.
2. Particularly in numbers, the number of functional illiterates in the U.S.—no matter how small the percentage—is a serious matter for concern.
3. Studies such as the APL attempt to define *functional literacies.* Such studies should become a starting point for refining categories and definitions relevant to this concern so that they are not subject to logical attacks such as Fisher's. This could then lead to iterative testing that might depict literacy trends.

OTHER TREND INDICATORS

There are other data over time that may be interpreted as reading trend indicators. Such information is often considered suspect by experimental researchers, but because nonexperimental methodology allows the researcher to report the real world instead of framing it in the laboratory, the data holds great promise both for defining literacy and determining its status. Furthermore, over time, it offers genuine indications of trends. Such research has its own standards which determine its quality and the dependability of its results. It involves much more than historical and survey techniques—including multi-perspective observation and interviewing. But to illustrate this promising point, one can cite some relatively simple statistics reported by the 1978 consumer research report of the Book Industry Study Group ([BISG] Cole & Gold, 1979). They point out rather eloquently that as reading is involved in literacy, ours is an increasingly literate society.

1. The number of libraries in the U.S. has tripled since World War II.
2. Between 1962 and 1974, the total number of print materials in U.S. libraries rose from 241 million to 397 million.
3. Eighty-three percent of Americans use libraries; 29% at least once every two months.
4. Growth in magazine circulation slowed from 37% in the 1940s to 6.3% in 1975; magazine circulation rose from 121 million to 255 million per issue; 3 out of 4 persons buy magazines.
5. The number of magazine titles increased from 219 in 1945 to 327 in 1975.
6. The 1800 newspapers published in 1976 equaled the same approximate number for 1940. There was a decrease of dailies in cities of one million or more population; an increase in those with 50,000–100,000.
7. Between 1950 and 1976, newspaper circulation rose from 54 million to 61 million.
8. Seventy-five percent of the population "reads" a newspaper every day.
9. In 1945, 5400 new book titles and 1200 new editions were published; in 1977, 27,400 new titles, 8100 new editions.
10. Yet annual unit sales of books decreased 1% between 1972 and 1976—from 1,259,600 to 1,243,700.
11. The number of establishments specializing in paperback books rose 2500% between 1958 and 1978.
12. In 1949, 25 to 30% of the population read at least one book a month and used a library regularly; by 1958, a different study showed that number down to 21%. The latter finding (Gallup) showed book reading at 21% in 1964 and at 26% in 1971.

The BISG study is rich with trend information that appears to have relevance to reading. It reports reading trends by types of books and by segments or ages of the population. It correlates more frequent reading and a higher number of books read to more education. It also contains significant data on many societal factors that have apparent relevance to reading in the United States. It is not the intent here to exemplify, beyond this single citation, the wealth of data that could be gathered on reading-related factors. Rather, the point is to illustrate briefly that achievement test data—even if they were complete—are but a single facet of a very complicated question.

CONCLUSIONS

In interpreting the total data pool summarized in this chapter, the caveats discussed qualify any conclusions about trends and stand themselves as conclusions with implications for future trend research. We must keep the following in mind:

1. The data do not support the claims that children today are poorer readers than those of previous generations. In fact, if one is concerned with basic literacy—as represented by the comprehension of everyday reading matter—the data tend to support the conclusion that today's children are better readers than children from any period in the past and that improvement in this area has been continuous in the history of education in the United States.

2. Most analyses of reading trend data suggest that reading performance in the lower elementary grades has improved continuously in the history of education in the United States and that in upper grades, where more sophisticated levels of comprehension are tested, there may be a decline in the gain pattern beginning in about the mid-1960s. Such conclusions, however, are subject to frequent sets of data which contradict these generalizations. The contradictions to the conclusion that performance at upper grade levels has declined include data from the more convincing trend studies conducted. Most conclusions need to be seriously qualified by factors such as age differences, dropout rate changes, and total population–sample changes.

3. Beyond the above generalizations, any trends suggested by the data vary from urban to rural areas, and between geographic regions of the country, socioeconomic levels, sexes, races, and countries of national origin.

4. Along a time line for any specific set of the factors just listed, performance scores do not increase or decline without interruption. While a general trend may be read in some data for some time spans, it is generally violated by the data for one or more individual years within the span. Frequently, when the data for these "erratic" years are compared to the data for the first or last year in the time span, the dif-

ference is greater than that between the first and the last year in the span. An example of this point can be seen in Table 5.20. If one is to claim a 1.9-month loss on the ITBS for fifth graders in Iowa between 1965 and 1975, one must overlook the data for 1966–1968; otherwise, the decline need be characterized as between 1967 and 1975 or, since the 1971 GE is but one-half month below that in 1965, as a decline beginning in 1971. In the fourth-grade data, it seems risky to talk about a .4-month decline between 1965 and 1975 when the 1975 GE is but .1 month behind 1971's and may be an atypical score in some broader time span. Note, too, that there was a decline of .7 months just between 1969 and 1970.

5. Any responsible comparison of achievement scores from different time periods must consider a host of factors that may operate on the groups tested and that may vary across time. Such factors involve educational, demographic, economic, and other societal factors which act as uncontrolled variables. Events, situations, and attitudes unique to time periods ought to be included and studied among such factors.

   Since such factors can interact in highly complex ways, most cannot be mathematically removed from the test scores and, so, must be weighed subjectively. While this should not inhibit recognition of a study's potential importance, conclusions and implications of any trend study are open to question and to interpretation from numerous perspectives. The achievement trend analysis which ignores such factors, however, is much more limited than one which takes them into account.

6. The purpose of tests inherently guarantees that test scores cannot produce absolute trends because:
   1. Differences in emphases and the content of specific test instruments limit the results of trend analyses that compare scores from more than one specific testing instrument.
   2. Factors unique to specific time periods influence curricula and the tests designed to measure them. Thus, using a test designed for one time period on subjects in a later time period tends to favor the earlier subjects and limit the implications of the comparison.

7. Changes in test scores do not necessarily relate meaningfully to changes in literacy. Tests vary considerably between publishers, editions, levels, and forms; and there is no scientific evidence that they are valid—either individually or as a group—as measurements of *literacy*, a term that has never been universally defined. Most tests are fairly narrow in their coverage; all are limited by questioning tech-

niques, such as multiple-choice; and we must accept the fact that measurement errors exist and that the tests are only estimates. Thus, even if a meta-analysis of the data gathered here could determine clear reading trends, its meaning in terms of life-competency would not be clear-cut. One way to improve the literacy score correlation of tests is to incorporate into score analyses other types of information relevant to literacy so as to broaden the attempt to assess how well schools are developing literacy.

8. Trend analyses should attempt to determine successes and failures in achieving broadly accepted educational goals. While "audit" studies of functional literacy conducted in a single time period may help determine literacy goals and even dictate needed changes in curricula, they cannot indicate whether schools are doing a better or poorer job than they did in time periods before the audit was conducted. If such audits were reiterated, they could indicate trends that are vulnerable to the same subject–instrument pitfalls inherent in achievement trend descriptions.

9. Little in the analysis of available reading achievement or literacy trend data justifies the national reaction of instituting competency testing. Some form of minimum competency testing is being used or considered in every state in the nation, and the rationale is clearly that some body of data shows that the schools are not doing their job. This is not to argue here that the minimum competency phenomenon is either bad or good—but rather to emphasize that it can hardly find justification in trend data.

Using any reasonable time span, the data tend to contradict the national alarm and to endorse rational analyses like Ralph Tyler's (Kiester, 1978):

> Many people really believe that in the past 100 percent of the population learned to read by the time they finished the primary grades. This was never true. When the country was founded, the best estimate is that 15 percent of the people were literate. They were plainsmen, frontiersmen, and farmers and they didn't need to read. By the time of the draft in World War I, about 45 percent of the 18-year olds were able to read. By World War II that got up to 65 percent. By the 1975 national assessment it was 80 percent, so we're moving along [p. 33].

REFERENCES

Bestor, A.E. *Educational wastelands.* Urbana: The University of Illinois Press, 1953.
Blomenberg, P. *A comparative study of selected socioeconomic and educational variables having probable influence on the reading achievement of sixth- and tenth-grade students*

in Indiana in 1944-1945 and 1976. Doctoral dissertation, School of Education, Indiana University, 1980.

Boss, M.E. Reading then and now. *School and Society,* 1940, *51,* 62–64.

Bradfield, R.H. Academic achievement: then and now. *Academic Therapy,* Summer, 1970, *4,* 259–265.

Brown, D. *Equating the 1972, 1974, 1975, and 1976 Educational Assessment Program Tests.* Occasional Paper in Education. Trenton, New Jersey: Bureau of Research and Assessment, Division of Research, Planning, and Evaluation, New Jersey State Department of Education, July, 1978.

Burke, N.F. and Anderson, K.E. A comparative study of 1939 and 1950 achievement test results in the Hawthorne School in Ottawa, Kansas. *Journal of Educational Research,* September 1953, *47,* 19–33.

Caldwell, O.W. and Courtis, S.A. *Then and now in education, 1845–1923.* Yonkers-on-Hudson, New York: World Book Co., 1924.

California Assessment Program. *Student achievement in California schools, 1977–78 annual report-reading: Grade 12* (excerpted). Sacramento: California State Department of Education, 1979.

California Assessment Program. *Survey of basic skills: Grade 6.* Sacramento: California State Department of Education, 1978.

Coffman, W. *Changes in average test scores: ITBS and ITED.* [Announcement] Department of Public Instruction, Iowa, September, 1979.

Cole, J.Y. and Gold. C.S. *Reading in America 1978.* Washington, D.C.: Library of Congress, 1979.

Cook, W.P. *Adult Literacy Education in the United States.* Newark, Delaware: International Reading Association, 1977.

Copperman, P. The achievement decline of the 1970's. *Phi Delta Kappan,* June, 1979, *60,* (10), 736–739.

Copperman, P. *The Literacy Hoax.* New York: Morrow, 1978.

Daughtry, B.H. *Are criticisms of modern schools justified?* Unpublished paper. University of Florida (at Gainsville), 1947.

Davis and Morgan. See Partlow.

Ebel, R. L. Declining scores: A conservative explanation. *Phi Delta Kappan,* December, 1976, *58*(4), 306–310.

Elementary reading test scores improve. *CAPtions.* Newsletter of the California Assessment Program, May 1979, *1*(1), 1.

Ellis, R.A. *The state of American reading habits and skills.* Unpublished paper. February, 1978.

Farr, R. *The teaching and learning of basic academic skills in schools: a testimony before the Senate Subcommittee on Education, Arts, and Humanities.* New York: Harcourt Brace Jovanovich, 1979.

Farr, R. and Blomenberg, P. Contrary to popular opinion. *Early Years.* May 1978–1979, *9*(9), 52–55, 68.

Farr, R., Fay, L., and Negley, H.H. *Then and now: reading achievement in Indiana (1944–1945 and 1976).* Bloomington, Indiana: Indiana University, School of Education, 1978.

Farr, R. and Olshavsky, J. Is minimum competency testing the appropriate solution to the SAT decline? *Phi Delta Kappan,* April 1980, *61*(8), 528–30.

Farr, R. and Tone, B. What does research show? *Today's Education,* November–December, 1978, *67*(4), 33–36.

Farr, R., Tuinman, J., and Rowls, M. *Reading achievement in the United States: Then and now.* Bloomington, Indiana: Indiana University, the Reading Program Center and the Institute for Child Study, August 1974.

Fay, L. An unpublished collection of reading assessment data from state departments of education. Indiana University, School of Education, Department of Reading (Bloomington, Indiana, 47405), 1978.

Finch, F.H. and Gillenwater, V.W. Reading achievement then and now. *Elementary School Journal,* 1949, *49,* 446-454.

Fish, L.J. Seventy-five years ago and today: comparison of exam and results, 1853-1928: entrance examination high school. *Education,* February 1930, *50,* 325-332.

Fisher, D.L. *Functional literacy and the schools.* Washington, D.C.: U.S. Dept. of HEW and National Institute of Education, January, 1978.

Flanagan, J.C. Changes in school levels of achievement: Project TALENT ten- and fifteen-year retests. *Educational Researcher,* September, 1976, *5,* 9-12.

Flanagan, J.C. and Jung. S.M. *Progress in education: A sample survey (1960-70).* Palo Alto, California: American Institutes for Research, 1971.

Flesch, R. *Why Johnny can't read.* New York: Harper and Row, 1955.

Flesch, R. Why Johnny still can't read. *Family Circle,* November, 1979, *93,* 26, 42-43.

Fridian, Sister M. Reading achievement in a Catholic parochial school. *School and Society,* November 1958, *86,* 403-405.

Gadway, C. and Wilson, H.G. *Functional literacy: basic reading performance.* Denver: NAEP, 1976.

Gates, A.I. *Attainment in elementary schools: 1957 and 1937.* New York: Bureau of Publications, Columbia Teachers College, 1961.

Gerberich, J.R. The first of the three R's. *Phi Delta Kappan,* March 1952, *33,* 345-349.

Golub-Smith, M. *Longitudinal trends in basic skills achievement in New Jersey, 1976-1978.* Occasional Papers in Education. Trenton, New Jersey: Division of Operations, Research, and Evaluation, New Jersey State Department of Education, April 1979.

Greider, W. Stop knocking our public schools. *The Washington Post,* June 3, 1979.

Harnischfeger, A. and Wiley, D.E. *Achievement test score decline: Do we need to worry?* Chicago: CEMREL, Inc., 1975.

Harnischfeger, A. and Wiley, D.E. Achievement test scores drop: So what? *Educational Researcher,* March 1976, *19,* 5-12.

Harnischfeger, A. and Wiley, D.E. The marrow of achievement test score declines. *Educational Technology,* June 1976, *16,* 5-14.

Harris, L. and Associates, *Survival literacy: Conducted for the National Reading Council.* New York: Author, 1970.

Harris, L. and Associates. *The 1971 National Reading Difficulty Index: a study of reading ability for the National Reading Center.* New York: Author, 1971.

Hodgkinson, H. What's right with education? *Phi Delta Kappan,* November, 1979, *61*(3), 159-162.

Howard, E.R. Positive side of test score debate cited. *School and University Review,* Fall 1979, *9*(4), 1-2.

Hunter, C. St. J. and Harman, D. *Adult literacy in the United States: a report to the Ford Foundation.* New York: McGraw-Hill, 1979.

Iowa Department of Public Instruction, *Final report of the Task Force on Student Achievement in Iowa.* Des Moines, Iowa: Department of Public Instruction, October, 1979.

Johnson, S. S. *Update on education.* Denver: The Education Commission of the States, 1975.

Kiester, E., Jr. Ralph Tyler: The educator's educator. *Change,* February 1978, *10*(2), 28-35.

Krugman, J. and Wrightstone, J.W. Reading: then and now. *Highpoints,* April 1945, *30,* 54-62.

Larsen, J.J., Tillman, C.E., and Cranney, A.G. Trends in college freshman reading ability. *Journal of Reading,* February 1976, *19,* 367-369.

Luther, G.H. *History of the first testing program administered in the year 1848 to pupils in the four senior schools of the Cleveland Public Schools: a comparison of the test results with the results of the same tests administered to 8A pupils in four junior high schools in October 1947* (Bulletin No. 38). Cleveland Board of Education, Bureau of Educational Research, 1948.

Michigan Department of Education. *Status report on Michigan basic skills education.* No date.

Miller, V.V. and Lanton, W.C. Reading achievement of school children—then and now. *Elementary English,* February 1956, *33,* 91–97.

Monteith, M.K. How well does the average American read? Some facts, figures and opinions. *Journal of Reading,* February 1980, *235*(5), 460–464.

Munday, L.A. Changing test scores: Basic skills development in 1977 compared with 1970. *Phi Delta Kappan,* May 1979, *60*(9), 670–671.

Munday, L.A. Changing test scores, especially since 1970. *Phi Delta Kappan,* March 1979, *607*(7), 496–499.

Murphy, R.T. *Adult Functional Reading Study.* Princeton, New Jersey: Educational Testing Service, 1973.

Murphy, R.T. *Adult Functional Reading Study.* Princeton, New Jersey: Educational Testing Service, 1975.

National Assessment of Educational Progress. Reading: Summary data (Report 02-R-00). Denver: NAEP, Education Commission of the States, July, 1974.

Northcutt, N. *Adult functional competency: A summary.* Austin: University of Texas (at Austin), March, 1975.

Partlow, H.R. *Arithmetic and reading yesterday and today.* Canada: Capp Clark Press, May 1955 (Gives data on Davis and Morgan study).

Riley, J.L. *The Springfield tests 1846-1905-6.* Springfield, Massachusetts: Holden Patent Book Cover Co., 1908.

Rogers, D.C. Success or failure in school. *American School Board Journal,* October 1946, 113.

Rogers, V.R. and Baron, J. Declining scores: A humanistic explanation. *Phi Delta Kappan,* December 1976, *58*(4), 311–313.

Silver, E.S. (Ed.). *Declining test scores: A conference report.* The National Institute of Education, U.S. Dept. of HEW, 1976.

Tiegs, E.W. A comparison of pupil achievement in the basic skills before and after 1945. *Growing Points in Educational Research.* Official report of 1949 meetings. Washington, D.C.: American Educational Research Association, National Education Association, 1949.

Tierney, R.J. and Lapp, D. (Eds.). *National Assessment of Educational Progress in Reading.* Newark, Delaware: International Reading Association, 1979.

Trace, A. *What Ivan knows that Johnny doesn't.* New York: Random House, 1961.

Tyler, L.E. Review in *The Seventh Mental Measurements Yearbook,* Oscar K. Buros (Ed.). Highland Park, New Jersey: Gryphon Press, 1972, pp. 667–668.

Walcutt, C.C. (Ed.). *Tomorrow's illiterates; the state of reading instruction today.* Boston: Little Brown, 1961.

Weinman, J.J. *Declining test scores: A state study.* Massachusetts Board of Education, no date.

Wirtz, W. *et al. On further examination: Report of the Advisory Panel on the Scholastic Aptitude Test score decline.* New York: College Entrance Examination Board, 1977. See also *Appendixes to On further examination.*

Witty, P. and Coomer, A. How successful is reading instruction today? *Elementary English,* December 1951, *28,* 451–478.

Wolf, R.M. *Achievement in America: National report of the United States for the International Educational Achievement Project.* New York: Columbia University, Teachers College, 1977.

Woods, E.L. Unpublished data from the Division of Psychology and Research, Public Schools of Los Angeles, California, 1935.

Worcester, D.A. and Kline, A. Reading achievement in Lincoln, Nebraska schools—1921 and 1947. Unpublished paper. Lincoln: University of Nebraska, 1947.

Wray, M. *A comparison of the reading achievement of sixth-grade students in 1917 and eighth-grade students in 1919 with sixth- and eighth-grade students in 1978.* Master's thesis, Bowling Green State University, 1978.

Yankelovich, Skelly, and White, Inc. *The 1978 consumer research study on reading and book purchasing.* Darien, Connecticut: The Book Industry Study Group, 1978.

Chapter 6

Trends in School Mathematics Performance

James T. Fey
Thomas Sonnabend

Public opinion surveys regularly confirm that mathematics is considered one of the most important topics in elementary and secondary school curricula. While this importance is satisfying to mathematics teachers, there is a price to be paid for the prestige. Educators, the press, and public constituents of education closely monitor the effectiveness with which school mathematics is being taught. For the last 10 years, reports seem to have been uniformly discouraging. A succession of titles like "Why Johnny Can't Add," "New Math Strikes Out," or "Consumers Flunk Math Test" suggest that student mathematics achievement is not what it should be. The list of conjectured causes for the situation is very long and full of contradictory assertions.

The criticisms of school mathematics have not gone unchallenged, but criticisms and rebuttals have both relied heavily on the personal opinions and limited observations of individual mathematics educators. Until very recently, national data on programs and student performance in mathematics have been gathered only sporadically and with limited scope. At the height of concern about mathematics programs in 1975, a National Advisory Committee on Mathematics Education (NACOME) tried to collect and analyze available evidence related to two questions: First, What do research and general achievement data say about the recent and current mathematics abilities of American young people? Second, What aspects of

The Rise and Fall of National Test Scores
Copyright © 1982 by Academic Press, Inc.
All rights of reproduction in any form reserved.
ISBN 0-12-068580-9

school curricula, instruction, and societal context most strongly influence achievement in school mathematics?

The answers that NACOME found, and evidence added since 1975, provide no simple cause and effect model describing mathematics in school. They do, however, enrich our understanding of how schools work and provide guides to more effective programs and program implementation. In this chapter we present an overview of the varied evidence on achievement in school mathematics, an analysis of factors influencing achievement, and a critique of the assessment tools and processes themselves.

WHAT'S UP? DOWN? THE SAME?

In 1975 NACOME concluded that in "pooling all recent reports of mathematics testing, we found an unmistakable trend of declining scores over the past ten years [Hill, 1976, p. 44]." While this seems a fair statement of the situation taken as a whole, the patterns of achievement are more varied than any single trend can accurately reflect.

College Entrance Test Results

The earliest and most prominently mentioned indicators of decline in school mathematics have been annual reports from the Scholastic Aptitude Test (SAT) and American College Test (ACT) progams. Most college-bound students take one of these two exams which are designed to help colleges predict their applicants' likely success in post-secondary schooling.

The SAT includes a mathematics section which intends to test (*a*)

TABLE 6.1
SAT Mathematics Scores, 1956-57 to 1978-79 (every other year given)

Year	National Mean	
1956-57	496	
1958-59	498	
1960-61	495	
1962-63	502	(one standard deviation
1964-65	496	approx. 115 points)
1966-67	495	
1968-69	491	
1970-71	487	
1972-73	481	
1974-75	473	
1976-77	471	
1978-79	467	

Source: Braswell, 1978, p. 174.

understanding of elementary and junior high school mathematics through algebra I; (*b*) applications of these concepts; and (*c*) nonroutine problem solving (Braswell, 1978).

Since 1962, the SAT mathematics mean scores have declined in every year except one, with a total drop of about one-third of a standard deviation. At the same time, there has been an even greater drop in SAT verbal mean scores, from 478 in 1962 to 434 in 1975. On the mathematics section, the number of high scores—those at 600 or above—dropped from 20.2% in 1962 to 16.4% in 1975 with a corresponding drop on the verbal section.

The ACT includes a mathematics usage test consisting of 40 items covering the curriculum through geometry and algebra II. This exam intends to emphasize "the solution of practical, quantitative problems . . . [stressing] reasoning in a quantitative context, rather than memorization [Ferguson and Schmeiser, 1978, p. 183]." Like the SAT national mean, ACT mathematics scores peaked in the early 1960s and have dropped about one-third of a standard deviation from that level to current levels. The decline appears to have been sharpest in the early 1970s, with nearly level scores for the past 6 years.

Two other interesting measures of the mathematical achievement of college-bound students are the College Board Level I and II achievement exams. Until very recently, the scores on these tests stayed relatively stable, in contrast to aptitude declines (Jones *et al.,* 1977).

Standardized Achievement Tests, K–12

For many years commercial test companies have marketed mathematics achievement batteries designed to help students, teachers, and schools measure their performance against national norms. Developers of these tests collect national samples of student performance to establish grade level and age expectations. At the same time a new test version is being normed, other students take the form being replaced in order to measure comparability of new and old test scores.

Since the early 1960s, the general pattern in these renorming assessments has been either stability or slight improvement in lower grade mathematics and sizeable drops in upper grades (Roderick, 1974; NACOME, 1975; Harnischfeger & Wiley, 1975; Munday, 1979). Standardized tests have been used as part of many state accountability testing programs since the late 1960s, with results generally confirming those of the test developers (NACOME, 1975). It is important to note, however, that declines in mathematics K–12 have almost always been accompanied by comparable declines in the reading and language skills performance of the same students.

Criterion-Referenced Tests

A norm-referenced standardized test establishes ranking of an individual student or group of students in relation to others of the same age or experience. The ranking is based on a modest sample of questions from broad content areas, and thus, student scores reveal little specific information about needed remediation or curriculum improvement. To get this kind of constructive feedback, many state assessments and local school systems have shifted to criterion-referenced testing programs tailored to measure student performance on specific skill and understanding objectives.

The most thorough national testing of this type is the National Assessment of Educational Progress (NAEP). The two completed NAEP mathematics assessments show levels of current performance in several content areas at several cognitive levels and they indicate patterns of change between the 1972–1973 and 1977–1978 school years. With a national sample of over 70,000 students at ages 9, 13, and 17, and an item pool of over 1000 exercises, NAEP results come closer to a national profile of school mathematics achievement today than any other data source. Two detailed interpretive articles in mathematics education journals (Carpenter *et al.,* 1980a, 1980b) describe current performance in number tasks, algebra, geometry, measurement, and statistics. The articles conclude that at all age levels, students "were successful on exercises at the knowledge and skill levels in many content areas. This is particularly true of whole number computation skills, knowledge of geometric terms, and some measurement skills." There were areas of unsatisfactory performance—notably percent, fractions, algebra, and area or volume—and "students demonstrated a lack of the most basic problem-solving skills [1980b, pp. 337–338]." The interpreters concluded that many students appear to be learning mathematics in a rote fashion with little understanding of the skills they are asked to perform.

Several of the significant changes in NAEP data from 1972–1973 to 1977–1978 reinforce the major conclusions just mentioned. Over 200 exercises from the 1973 assessment were repeated in 1978. Table 6.2 shows the changes in average performance on those items at each age level. While all three age levels experienced decline in performance, the specific location of change is available in NAEP data reports.

At all three age levels, there was no change in average performance on mathematical knowledge items (those items calling for recall of specific facts). On items calling for arithmetic or algebraic computation, average performance of 9-year-olds was the same in both years, average 13-year-old performance dropped 2%, and average 17-year-old performance dropped 5%. The decline among 13- and 17-year-olds was concentrated in fractions,

TABLE 6.2
Average Performance on Mathematics Items – National Assessment
of Educational Progress, 1973-1978

	Number of Items	Percent Correct	
		1973	1978
Age 9	55	38	37
Age 13	77	53	51
Age 17	102	52	48

Source: National Assessment of Educational Progress, 1979,
p. 2.

percents, exponents, English system measurement, and algebraic equations. None of the groups decreased in ability to work with whole number computation, metric measurement, or graphing.

The pattern of change in performance on mathematical understanding items was similar to that for skills, but the greatest overall decline in the three age levels was on problem solving. At all age levels, the percents correct are disappointingly low and falling. Students seemed to have particular trouble with multi-step problems calling for complex analytical thinking (Table 6.3).

These NAEP mathematics data confirm longer trends of declining achievement, particularly in upper grades and secondary school. But the detailed description of varying achievement patterns and trends in specific topic areas suggests that a simple causal explanation is unlikely to be found.

TABLE 6.3
Average Performance on Mathematics Applications – National
Assessment of Educational Progress, 1973-1978

	Number of Items	Percent Correct	
		1973	1978
Age 9	9	38	32
Age 13	12	42	38
Age 17	25	33	29

Source: National Assessment of Educational Progress, 1979,
p. 12.

WHAT IS CAUSING ACHIEVEMENT DECLINES?

The 15-year decline of school mathematics test scores has been a continuing embarrassment and source of frustration to mathematics educators. During the 1960s, teachers, mathematicians, and curriculum developers devoted immense energy to the updating of existing school mathematics programs and the reeducation of teachers. Critics point to the test scores and argue that those reform efforts were misguided and unsuccessful. Defenders of the "new math" initiatives respond that the ideas were good, but successful implementation was stymied by conservatism in schools and by societal problems that overwhelm the academic goals of education.

The search for explanations of unsatisfactory classroom results has been characterized primarily by the sharing of personal impressions and opinions, but a growing body of research evidence permits more informed assessment of the various conjectures.

Influence of Social Context of Education

The test score news in mathematics is discouraging, but mathematics educators take some solace from the concomitant declines in nearly every other school subject area. This broad pattern of falling academic performance suggests that changes in societal attitudes toward and in support of education are most influential factors. The list of potential culprits is long: time spent watching television and not doing homework, broken homes or homes with two parents working and little time to devote to children, lower general societal regard for academic achievement (particularly in sciences during the 1970s), and so on.

There has been little research exploring the impact of such societal factors on mathematics. A Pennsylvania study (Guerriero, 1979) did find that television viewing time had a slight negative correlation $(-.15)$ with school mathematics achievement among eighth graders. But there is little other evidence or, for that matter, professional discussion of ways that television viewing might cause specific improvements or declines in school mathematics scores.

The conjecture that mathematics performance has declined because of diminished societal regard for and support of sciences faces contradictory evidence. A 1978 National Science Foundation (NSF) survey of the status of school mathematics, science, and social science revealed very strong professional and public support for school mathematics (Dawson, 1978; Fey, 1979a, 1979b). This high regard for mathematics is reflected in school time allocated to the subject. At the elementary level, it appears that mathematics gets a substantial share of instructional time (see Table 6.4).

TABLE 6.4
Time Devoted to Mathematics, Science, Social Studies, and
Reading in Elementary School (In minutes per day)

	Grade Range	
	K-3	4-6
Mathematics	38	44
Science	19	35
Social Studies	22	40
Reading	95	66

Source: Weiss, 1978, pp. 50-51.

Mathematics educators interpreting these data have seen a more subtle pattern that they think explains the test score trends of the past 10 years. They argue that while the public supports mathematics as a basic skill or minimal competence, and while teachers in elementary school reflect this concern in time allocation, the type of mathematics emphasized is arithmetic computation. In moving "back to basics," schools have, for the last 6-8 years, concentrated on building computational skills, the focus of lower-grade curricula. At the same time, however, there is evidence that students in senior high school are not enrolling in challenging mathematics courses. The sharp drop in advanced algebra (see Table 6.5) and senior mathematics enrollments is countered somewhat by the increase in calculus. But, on the other hand, there are suspicions that the increase in elementary algebra is a misleading statistic resulting from many students choosing to spread the traditional ninth grade course over 2 years. Data on college level enrollments 1970-1975 show a similarly sharp drop in numbers of students continuing mathematics study into advanced courses. Instead,

TABLE 6.5
Enrollment in Selected Secondary Mathematics Courses 1973-1977
(in thousands)

	1972-73[a]	1976-77[b]
Business Mathematics 9-12	392	609
Elementary Algebra	2052	2825
Advanced Algebra	1808	1317
Advanced Senior Mathematics	259	225
Calculus	55	105

[a]Source: Osterndorf, 1975, p. 10.
[b]Source: Weiss, 1978, p. 59.

more students are taking practical engineering, computer, and business courses (Fey, Albers, & Jewett, 1976).

These data seem helpful in explaining the SAT–ACT score declines. But there are, of course, a variety of other conjectures about that steady slide—many focusing on the broadened school population seeking college admission. These issues are discussed elsewhere in this volume.

Putting the separate pieces of data together, one could argue that the general pattern of the 1970s—steady or improving elementary grade test scores and declining secondary grade performance—reflects student educational priorities and curricular choices. But this explanation does not easily extend to the earlier period, 1963–1970, when Sputnik-inspired national priorities and school mathematics enrollments favored the most advanced courses.

The NSF status surveys (Stake & Easley, 1978; Weiss, 1978) gave some additional insight into the possible impact of education's social context on student achievement. A questionnaire survey of several thousand mathematics teachers asked the teachers to rate the seriousness of various potential problems. Eighty percent of secondary-school teachers said that low student interest in mathematics is a problem (Weiss, 1978, p. B-126). When other teachers were interviewed in case studies, they repeatedly mentioned the difficulty of motivating students. Remarks like "I find it very difficult to play Johnny Carson everyday [M. L. Smith 1978, pp. 2-9]," or "kids are lazy . . . I think they should be required to take a little more. The program is getting watered down [Steffensen, 1978, pp. 12-35]," reflect widespread perceptions of student attitudes. A student comment supported these views: "I haven't heard anyone, anyone except a mathematics teacher say that math is great. You gotta know how to figure your income tax, how to get money from banks, know to buy or sell stuff, . . . maybe a couple other things and that is it [Denny, 1978, pp. 1-110]."

In summary, there is some evidence, from studies directly focused on school mathematics, that recent achievement trends reflect changes in the broader context of education itself. Some educators assert that turmoil in the school as a social organization, diminished respect for authority, and instability of families overwhelm the efforts of academic instruction in all areas. These might be accurate insights, but others argue that current curricular and instructional practices within mathematics share equal blame for discouraging test scores.

Influence of Curricular Changes

School mathematics was one of the central participants in the well known post-Sputnik decade of curricular innovation, so declines in student achievement have frequently been interpreted as consequences of poor curriculum development. The basic lines of that criticism were most forcefully and eloquently argued by mathematician Morris Kline in a series of talks and papers and in the 1973 book, *Why Johnny Can't Add*. Throughout the opposition to "new math" goals and programs, Kline contended that the new content (set theory, logic, abstract algebra, etc.) was inappropriate for school mathematics; he faulted rigorous deductive presentation of ideas and excesses of terminology and symbolism; and he criticized inadequate attention to the applied scientific origins of and motivation for school mathematics topics.

The 1975 NACOME analysis addressed each of those charges, asking, "Are the claims accurate descriptions of school programs in operation?" and "Do the curricular initiatives explain correlated test score declines?" The details of that analysis are available in the NACOME report. However, the central points in the committee's conclusions are: (*a*) There were, undoubtedly, curricula that poorly incorporated modern ideas, and there were teachers who failed to see the distinction between form and substance as they introduced the new ideas. However, many of the criticisms leveled at "new math" curricula address only caricatures of the actual programs developed by projects like the School Mathematics Study Group (SMSG). (*b*) Even if those programs were flawed, evidence suggests that all the innovation efforts made only a modest impact on the majority of schools and teachers. Thus, it would be inaccurate to attribute test score declines to curricular change.

The NACOME conclusions drew on several sources of data. Firstly, as soon as experimental curricula entered schools for pilot testing in the early 1960s, small-scale investigations comparing new and traditional programs began. These studies usually showed no major difference between programs when scores on conventional standardized tests were the criterion measures. Sometimes the students using conventional textbooks were slightly better at computation, but when the test stressed terminology and concepts unique to the new programs, the modern textbooks were more effective.

To investigate more thoroughly the connections between textbooks or curriculum and student achievement, SMSG began a 5-year National Longitudinal Study of Mathematical Abilities (NLSMA) in 1962. The NLSMA study set out to describe the patterns of achievement associated with the use of different kinds of textbooks. After collecting data on performance of over 100,000 students using many different textbooks, NLSMA was led to the following conclusions:

1. At grades 4, 5, and 6, students in SMSG programs scored below the grand mean on 9 out of 11 computation scales, but they were above the grand mean on all 19 comprehension scales (and were higher than all other text users on 14 of the 19 comprehension scales).

2. At grades 7, 8, and 9, students who used a conventional text did relatively well on computation tests and relatively poorly on more complex items. The SMSG group fit the opposite pattern.

3. In grades 10, 11, and 12, there were fewer differences between curricula (data available to authors through NACOME report and files). The NLSMA data seem to support the apparently obvious idea that curricula do influence student achievement patterns and that at least one modern program was successful in promoting comprehension and problem-solving goals (though there was often striking difference between the achievement profiles of the various modern textbooks).

The NACOME conclusions about limited implementation of new curricula rest on other data sources. In 1974, the National Council of Teachers of Mathematics (NCTM) supported a questionnaire survey of instructional practices in third and fifth grade mathematics classes. The results led interpreters to conclude that "almost none of the concepts, methods, or big ideas of modern mathematics programs have appeared in [the average] classroom [Price, Kelley, & Kelley, 1977]."

This finding was largely confirmed by the 1977–1978 NSF status surveys. Those data revealed that prior to 1976–1977 only 30% of the sampled school districts had used at least one of the federally funded innovative mathematics curricula. Only 9% were still using those materials in 1976–1977, though both figures must be increased by some factor that accounts for the translation of new ideas into commercial texts (Weiss, 1978, p. 79).

The NSF surveys also revealed that many teachers who might have flirted with innovative mathematics programs are now heading back to basics. One teacher commented, "The brass tacks are learning addition and subtraction. That's it." Another said, "You might as well forget about teaching conceptual mathematics to 75% of the children in elementary school [Denny, 1978, pp. 1–31 through 1–34]."

The argument against curricular change as the central agent in test score decline is that massive change never really took place and, where some change had, there were programs of demonstrated effectiveness available. Before looking elsewhere for villains, it is important to point out that many still feel that "new math" has been a factor in the *achievement* slide, and there are some data to support the contention.

For instance, the state of New Hampshire began statewide testing of grade 8 students as early as 1957. When the seventh consecutive testing program revealed a sudden drop in arithmetic computation, research was initiated to assess the impact of curricular change on this decline. From 1965 to 1967, Austin and Prevost (1972) followed the performance of eighth graders in traditional, transitional, and modern programs. They found steady declines in computational ability of all three groups—the statewide mean slipped from a grade equivalent of 8.8 in 1963 to one of 7.8 in 1966 and 6.8 in 1967. The comparison of curricular effects was not very clear or simple. In 1965, the modern program group was outperformed by both of the other groups; in 1967, the reverse was true. A 1967 follow-up of 1965 eighth graders found that those with the modern background did significantly better than traditionally prepared students on tests of algebra and geometry.

The fact remains that in New Hampshire, as in many other states, declines in mathematics computation scores coincided with national efforts to redirect curricular priorities toward more conceptual programs. The continuation of this slide in the 1970s seems to call for a different explanation, and many have looked to changes in teachers and instructional style in search of adequate explanations for this period of decline.

Influence of Teachers and Instructional Style

Those who see changes in teachers and instructional style as factors in the 1970s test score decline might argue that throughout the 1970s, schools have emphasized individualization of instruction: specification of curricula in discrete behavioral objectives, instruction for students through self-study learning packets, and pass/fail assessments of student mastery in each objective. Research shows that individualized programs have had only limited success (Schoen, 1976) and that teachers who spend considerable time on expository instruction are more effective than those who utilize much student seatwork (Evertson et al., 1980; Good & Grouws, 1977).

The problem with this argument is the extensive data showing that mathematics instruction at all levels has changed very little in the past twenty years (Price, Kelley, & Kelley 1977; Weiss, 1978; Stake et al., 1978). The typical pattern is a routine of going over the previous day's assignment, teachers telling how to do the next assignment, and student opportunity to begin work on the new assignment. If anything, teachers have recently pulled back from whatever instructional innovation they might have tried during the late 1960s and early 1970s. In the NSF status survey, one high school teacher expressed the thoughts of many colleagues: "We've found that traditional methods work. This is the way it was taught to us in high school and . . . college and the way it works for us [M. L. Smith, 1978, pp. 2–11]."

If there is a way that teachers seem to have changed, it is in their attitudes toward their students and their jobs. Observing the "burnout" phenomenon, one of the NSF survey case study authors described the prob-

lem as "a flatness, a lack of vitality, a seeming lack of interest in the curriculum by both the teacher and the children, a lack of creativity and curricular risk taking, a negativism toward the children—they're spoiled, they
don't care, they don't try— and sometimes a negativism toward colleagues,
administrators, and college and university training programs [L. M. Smith,
1978, pp. 3-84]."

This disillusion among teachers and students has, undoubtedly, complex
origins and consequences, but its potential impact on student achievement
is all too plausible. A downward spiral of test scores, frustration, and
disappointment—often accelerated by antagonistic public scrutiny—is evident in nearly every discussion of school mathematics today.

While offering insight into explanations of the recent decline in school
achievement, the picture of discouragement and disillusion in no way fits
the ambitious, optimistic 1960s. For at least 15 years the test score indicators of performance in school mathematics have declined, but the factors which have contributed to that decline are numerous, interacting, and
varying within the broader school and social context of mathematics instruction.

ARE THE TESTS VALID AND
RELIABLE INDICATORS?

In the early 1960s, when curriculum innovators sought to compare new
and traditional program effects, the natural question was, "Which test will
be fairer, the existing standardized tests, based on the old syllabus, or new
tests, reflecting new goals?" Concern over validity of achievement tests and
proper interpretation of results has continued in vigorous debate to the
present.

Recent conferences and research studies have examined the norm-basing
of standardized tests, the influence of factors unrelated to mathematics
ability, and the inability of standardized tests to evaluate problem-solving
skills. Mathematics educators also question the usefulness of test results in
selecting curricula and in evaluating the relative standing of schools, the effectiveness of teachers, and the specific strengths and weaknesses of
students.

Problems with Norm-Referenced Testing

One of the main targets in the criticism of testing practice has been the
common norm-referenced test. In 1975, the NACOME report identified
several crucial flaws in standard usage.

First, norm-referenced tests are used to compare individual test takers and schools to national norming populations. Since these exams employ items that test content common to most schools, they ignore the unique aspects of the curriculum in a given school.

Second, norm-based test results are frequently reported using grade-level scores. For instance, a score of 5.8 would mean performance comparable to that of the average student who is 8 months into the fifth grade. NACOME pointed out the drawbacks of such scoring. "It does not have a statistical interpretation in terms of the population of students at the grade level of interest . . . and does not explicitly identify [a student's] position in the distribution of scores for his age group peers [NACOME, 1975, p. 122]." Most people do not understand that half of the students taking a standardized test should be expected to score below grade level since that score is the average. Many interpret any score below grade level as unsatisfactory. In addition, scores of one year or more above or below grade level are misleading. "A bright third grader who receives a score of 5.5 is not necessarily performing like a mid-year fifth grader." His score indicates that he did very well on the third grade test items, but he "might be able to do very few of the fifth-grade level problems [NACOME, 1975, p. 123]."

Because of these flaws in grade level score reporting, NACOME recommended that grade level scoring be abandoned. As a viable alternative for score reporting they proposed the use of *stanines* which give a score as one of the integers from 1 through 9. Stanines indicate where the student's score falls within his peer group population, and are less likely to be misinterpreted since they do not make the fine distinctions of percentile and grade level scores (NACOME, 1975, p. 123).

Problems with Extraneous Test Score Influences

Many factors unrelated to mathematics ability significantly affect standardized test scores. For instance, recent studies indicate that test-taking skills affect test results. Eakins (1976) found that first graders improved an average of 2.8 months on the Metropolitan Achievement Test after multiple presentations of a test taking unit. Ginther (1978) noted that "the ability tests that are good predictors of mathematics achievement for Anglo students are not good predictors for Chicano students [p. 118]." In her study, Ginther found that pretraining seventh-grade Chicanos for an arithmetic reasoning test improved the reliability of the test, while the pretraining of Anglos had no effect on the reliability of the test for them.

The NACOME report describes unique qualities of the standardized testing situation which have unpredictable effects on individuals. The standardized test setting creates a competitive, high-pressure situation because of its special seating arrangements, special answer sheets, timing of the test, and long hours of testing.

Any classroom teacher who has administered these tests is familiar with the variety of reactions students have. Some students react to the pressure by becoming very tense and unable to concentrate. Many younger students become restless after a short period of time. Some older students will deal with the pressure by acting apathetic about the situation. A significant number of students are adversely affected by the test-taking situation, and the validity of their test results is questionable.

NACOME (1975) recommended that special test arrangements be eliminated wherever possible. They suggested that special seating, special test days, and long hours of testing be eliminated and that shorter sections be administered on different days in the regular classroom setting (p. 128).

Research suggests that the use of special test booklets and separate answer sheets has a significant effect on test scores. Garigliano (1975) examined the test papers of 51 fourth-grade students on the mathematics portion of the Stanford Achievement Test. He found 38 answers that were correct on scrap paper but incorrectly transferred to the answer sheet. Nineteen problems were copied incorrectly from the test booklet, and thus, the wrong problem was done correctly on scrap paper. On the average, each student missed about one item as a result of a clerical error in this transfer process. A drop of one raw score point is usually equivalent to a drop of .3–.9 of a grade level.

Gregory (1979) gave a 33-item arithmetic test to 842 students in grades 6–9. The items involved nothing more difficult than whole number addition and subtraction problems with answer choices numbered from 1–5. Using a separate answer sheet, the 842 students made a total of 1255 errors. A week later, 719 students retook the test and were allowed to mark their answers in the test booklet. They made a total of only 66 errors! "Of the 342 students in the new sample who had missed one or more items with the answer sheet, 296 got perfect scores when they marked answers directly on the test [Gregory, 1979, p. 51]." In both the Garigliano and Gregory studies, the use of a separate answer sheet resulted in a significant decline in some students' scores.

Garigliano's research also suggests that the time of year a standardized test is given has a significant effect on test results. He gave the Stanford Achievement Test in mathematics to 124 fourth graders in October and in May. In October, 55% scored below grade level (4.1). In May of the following year, only 35% scored below grade level (4.8).

Garigliano also examined the effects of timing fourth graders on the mathematics portion of the Stanford Achievement Test. He argued,

"Achievement tests should be a test of power, not one of time [p. 410]" and found that students taking the timed test had no time to attempt 348 items; students taking an "almost unlimited" amount of time on the test reached all but five items.

Two hundred fourteen of Garigliano's students were given a timed test in October. In May of the following year, they were divided into two comparable groups. The May timed group improved by 1 grade equivalent while the May untimed group improved their October scores by an average of 1.6 grade levels. Both studies suggest that the timing of tests makes a significant difference. As NACOME pointed out, "it is not so clear that speed is of major importance in mathematical understanding . . . a much better indication is the ability to solve difficult and complex mathematical problems [1975, p. 127]."

Problems with the Scope of Standardized Tests

In recent years, as mathematics educators have focused attention on problem solving and interdisciplinary curricula, they have criticized standardized multiple choice tests for examining only a limited portion of what students should be learning. Standardized testing calls for skill performance or problem analysis at a rate of about one problem per minute. Many mathematical problems in the "real world" are undoubtedly of this type, but most realistic problems are more complex. In addition to recommending the development of better ways to measure this higher level behavior, test critics have called for test formats that require the processes used by students in solving problems to be demonstrated. Such information would, of course, be useful in diagnosis and remediation of individual strengths and weaknesses.

Problems in the Use of Test Results

Closely related to the issue of test validity is the proper use of standardized test results to evaluate students, teachers, and schools. Responding to perceptions that test results, with all their limitations in measuring student knowledge, are being used more and more as the sole criteria for evaluating student and program accomplishments, the recent NCTM *Agenda for Action* called for a reversal of this tendency to rely so heavily on test scores. The NCTM (1980) document based its objection to current test practice on the contention that "It is imperative that the goals of a mathematics program dictate the nature of the evaluations needed to assess program effectiveness, student learning, teacher performance, or the quality of materials. Too often the reverse is true; the tests dictate the programs [p. 14]."

This NCTM position also reflects widespread concern that the test score pressure has led many schools and teachers to "teach for the test." Every teacher has faced the popular question, "Do we have to know this for the test?" Mathematics educators are arguing vigorously that it would be unfortunate if this question came to dominate every aspect of curriculum-making and instruction.

In summary, the test score controversy has provoked thorough examination of the validity and reliablity of current measurement instruments and procedures. Mathematics educators have questioned norm-referencing; grade level score reporting; and the significant effects of factors unrelated to mathematical ability, such as test-taking ability, cultural background, multiple choice format, separate answer sheets, timing of tests, and the time of school year when the test is given. There is also concern about the limitations of standardized tests for assessing higher level cognitive abilities and about the appropriateness of using standardized test scores as a measure of individual achievement or program success.

CONCLUSIONS

In the national concern about declining performance of elementary and secondary school students, a great deal of attention has focused on mathematics test scores. These data show an overall pattern of decline for the past 15 years. However, the decline has not been uniform across grade levels or across the topic areas within mathematics. Furthermore, many of the common explanations for the decline have limited validity, at best.

Mathematics scores of students seeking college admission peaked in the early 1960s and have declined steadily since. Elementary and secondary school data from standardized tests and national assessment indicate steady or slightly increasing performance among the youngest students, but continued decline at junior high and senior high school levels. However, the pattern of decline has changed over the years. Through the 1960s, the greatest decline occurred in the area of computational skills; during the 1970s, comprehension and problem solving have declined.

In the search for cause and effect explanations for the generally poorer test performance of students, educators and concerned public commentators have speculated on a wide variety of the influential curricular, instructional, and school–society environmental factors. Unfortunately, there is evidence which supports and contradicts each proposed explanation. For instance, declines during the 1960s were concurrent with major attempts to change school mathematics curricula, but recent evidence suggests that those presumably damaging curricula changes never really influenced the programs of very many U.S. schools. Declines during the 1970s have been accompanied both by major efforts to equalize educational opportunity and by a general societal disenchantment with goals and methods of science, but recent evidence also suggests that mathematics has not shared in the diminished regard for science and that performance of school children in mathematics has declined among all socioeconomic groups.

Despite the desire of some school critics to find simple, easily repaired flaws in mathematics education, it seems likely that the general decline of mathematics achievement has been caused by a complex of factors, each playing a more or less influential role at given times and in given situations. Attempted curricular innovation—and the associated misinterpretations and lack of goal clarity—during the 1960s was very likely a factor in the declines of that era. But it is also likely that college entrance scores, for example, were depressed by the natural influence of many less able students seeking access to higher education. During the 1970s, schools were the focus of a great deal of social turmoil that undoubtedly affected academic performance, but it is also likely that reactionary curricular trends of the "back to basics" variety have influenced the trend toward poorer problem-solving performance.

For those who have been active in mathematics education during the tumultuous 2-decade period from 1960–1980, reflection on the successes and failures holds a number of very valuable lessons. First, the "new math" experience of the 1960s made everyone aware of the great difficulties involved in changing school programs. It is now very clear that effective change requires careful preparation—demonstration research, consultation with affected parties, teacher education, and adjustment in the expectations of evaluation bodies. Second, those who attempted change during that period have acquired a much deeper appreciation for the social context of all academic efforts in public education. Mathematicians proposing radical change in the content and organization of school curricula naively assumed that a logical presentation of the structure of their discipline would entice students and promote effective learning. Recent experience has shown the limitations of this viewpoint; that is, the importance of matching curricula to psychological development of young people and to the social values and interests that students bring to mathematics class for their one hour each day.

Among all the conclusions that can be drawn from recent experiences in mathematics education, perhaps the least appreciated outside the profession is the fundamental difficulty in assessing the impact of schooling. More than any other country in the world, the United States measures the effects of education by scores written on paper with pencils. Some very thoughtful analyses of common testing practices have demonstrated the major flaws in practices such as grade level score reporting and mass testing in standardized, time-bound, battery formats. The limitations of multiple-choice testing for the assessment of growth in problem solving and analytical thinking are in no way reflected by the standardized tests which the nation commonly uses to evaluate school progress.

There are unmistakable signs that school mathematics achievement has declined in degree during the past 15 years. However, careful study of available evidence should suggest to anyone that solutions lie in positive action to improve the many factors contributing to the unsatisfactory situation, not in the type of hit-and-run charges that so often fill the air upon announcement of each new piece of test data.

REFERENCES

Austin, G. & Prevost, F. Longitudinal evaluation of mathematical computational abilities of New Hampshire's eighth and tenth graders, 1963-1967. *Journal for Research in Mathematics Education,* 1972, *3,* 59-64.

Braswell, J. S. The College Board Scholastic Aptitude test: An overview of the mathematical portion. *Mathematics Teacher,* March, 1978, *72,* 168-181.

Carpenter, T. P., Kepner, H., Corbitt, M. K., Lindquist, M. M., and Reys, R. E. Results and implications of the second NAEP mathematics assessment: Elementary school. *The Arithmetic Teacher,* April, 1980a, *27,* 10-12, 44-47.

Carpenter, T. P., Corbitt, M. K., Kepner, H. S. Jr., Lindquist, M. M., and Reys, R. E. Results of the second NAEP mathematics assessment. *The Mathematics Teacher,* May, 1980b, *73,* 329-338.

Coburn, T. G., Statewide assessment and curriculum planning: One state's experience. *Arithmetic Teacher,* November, 1979, *26,* 14-21.

College Entrance Examination Board. *National report—college bound seniors, 1979.* Princeton: College Entrance Examination Board, 1979.

Dawson, E. K. Survey findings and corroborations. In Stake and Easley (Eds.), *Case studies in science education,* 1978, 18-1 to 18-115.

Denny, T. River acres. In Stake and Easley (Eds.), *Case studies in science education,* 1978, 1-1 to 1-125.

Eakins, D. *et al.* Effect of an instructional test-taking unit on achievement test scores. *Journal of educational research,* November, 1976, *70,* 67-71.

Evertson, C. M., Emmer, E. T., and Brophy, J. E. Predictors of effective teaching in junior high mathematics classrooms. *Journal for Research in Mathematics Education,* May, 1980, *11,* 167-178.

Ferguson, R. L. & Schmeiser, C. B. The mathematics usage test of the ACT assessment program: An overview of its purpose, content, and use. *Mathematics Teacher,* March, 1978, *71,* 182-192.

Fey, J. T., Albers, D. J., and Jewett, J. *Undergraduate mathematical sciences in universities, four-year colleges, and two-year colleges, 1975-1976.* Washington, D. C.: Conference Board of the Mathematical Sciences, 1976.

Fey, J. T., Mathematics teaching today: Perspectives from three national surveys. *The Arithmetic Teacher,* October, 1979a, *27,* 10-14.

Fey, J. T. Mathematics teaching today: Perspectives from three national surveys. *The Mathematics Teacher,* October, 1979b, *72,* 490–504.

Garigliano, L. J. Arithmetic computation scores. *School Science and Mathematics.* 1975, 399–412.

Ginther, J. R. Pretraining Chicanos before administration of mathematics predictor test. *Journal for Research in Mathematics Education,* March, 1978, *9,* 118–125.

Good, T. & Grouws, D. Teaching effects: A process–product study in fourth grade mathematics classrooms. *Journal of Teacher Education,* 1977, *28,* 49–54.

Gregory, J. W. Test failure and mathematics failure—there is a difference. *Arithmetic Teacher,* November, 1979, *26,* 50–52.

Guerriero, C. A. Predictors of eighth grade mathematics achievement. *Focus on Learning Problems in Mathematics,* July, 1979, *1,* 69–73.

Harnischfeger, A. and Wiley, D. E. *Achievement test score decline: Do we need to worry?* Chicago: CEMREL, Inc., 1975.

Hill, S. Issues from the NACOME report. *The Mathematics Teacher,* October, 1976, *69,* 440–446.

Jones, C., Rowen M. R., and Taylor, H. E. The CEEB mathematics achievement tests. *The Mathematics Teacher,* March, 1977, *70,* 197–208.

Kline, M. *Why Johnny can't add.* New York: St. Martin's Press, 1973.

Lipsitz, L. (Ed.). *The test score decline.* Englewood Cliffs, N. J.: Educational Technology Publications, 1977.

Munday, L. A. Changing test scores: Basic skills development in 1977 compared with 1970. *Phi Delta Kappan,* May, 1979, *60,* 670–671.

National Advisory Committee on Mathematics Education (NACOME). *Overview and analysis of school mathematics grades K–12.* Washington, D.C.: Conference Board of the Mathematical Sciences, 1975.

National Assessment of Educational Progress. *Changes in mathematical achievement, 1973–1978.* Denver: Education Commission of the States, 1979.

National Council of Teachers of Mathematics. *An agenda for action: Recommendations for school mathematics of the 1980s.* Reston, Virginia: NCTM, 1980.

Osterndorf, L. *Summary of offerings and enrollments in public secondary schools, 1972-1973.* Washington, D.C.: National Center for Education Statistics, 1975.

Price, J., Kelley, J. L., and Kelley, J. 'New math' implementation: A look inside the classroom. *Journal for Research in Mathematics Education,* November, 1977, *8,* 323–331.

Roderick, S. A comparative study of mathematics achievement by sixth graders and eighth graders, 1936 to 1973, 1951–55 to 1973 and 1965 to 1973. doctoral dissertation, University of Iowa, 1974. Dissertation Abstracts International, March, 1974, 34A, 5601–5602.

Schoen, H. Self-paced instruction in mathematics instruction: How effective has it been in secondary and post-secondary schools? *The Mathematics Teacher,* May, 1976, *69,* 352–357.

Smith, L. M. Alte. In Stake and Easley (Eds.), *Case studies in science education,* 1978, 3–1 to 3–136.

Smith, M. L. Fall river. In Stake and Easley (Eds.), *Case studies in science education,* 1978, 2–1 to 2–23.

Stake, R. E. and Easley, J. (Eds.). *Case studies in science education,* Washington D.C.: U.S. Government Printing Office, 1978.

Steffensen, M. The various aims of science education. In Stake and Easley (Eds.), *Case studies in science education,* 1978, 12–1 to 12–45.

Weiss, I. *Report of the 1977 national survey of science, mathematics, and social studies education.* Washington, D.C.: U. S. Government Printing Office, 1978.

Chapter 7

National Science Test Scores: Positive and Negative Trends in Academic Achievement in Science with Particular Emphasis on the Effects of Recent Curriculum Revision

Henry H. Walbesser
Cheryl Gonce-Winder

INTRODUCTION

There is no question that this country has come to prize science and its technology very highly. There is the continuous expectation that an entire set of medical, social, political, and economic handicaps will be remedied through increased scientific development. The visual media—film and television particularly—provide a constant flow of science heroes and heroines whose glorious efforts improve humankind's condition, and science fiction literature abounds on the bookstore shelves.

Positive attitudes toward and interest in science (Welch, 1980; National Assessment of Educational Progress, Oct. 1979), however, do not appear to assure high levels of performance among elementary and secondary school science students. The three national assessments conducted by the National Assessment of Educational Progress (NAEP) between 1969 and 1977 clearly indicate score declines across all three of the tested age groups (9, 13, 17). What explanations might there be for this discrepancy between interest and performance?

Conclusions which focus on the notion that insufficient attention has been given to the science curriculum are hasty at best, and generally uninformed about classroom practice. A review of the federal government's investment in elementary and secondary science curriculum development cites the millions of dollars and thousands of the best scientific minds that were

163

Copyright © 1982 by Academic Press, Inc.
All rights of reproduction in any form reserved.
ISBN 0-12-068580-9

used to effect more sophistication and relevance in instruction (Elementary school: S-APA, SCIS, ESS; Middle school: ESCP, ISCS; High school: BSCS, CBA, Chem Study, PSSC, Harvard Project Physics. Most of these programs began in the early 1960s). Such efforts saw science education introduced as an integral part of elementary instruction. The life, physical, and earth sciences became more definitively established in the middle school program, and more active student participation in the process of science became evident at all educational levels. Textbooks published after the reform wave illustrate this characteristic more than any other. At the same time, local school systems' budgets were likewise reflecting the curriculum shift with allotments for laboratory costs that began to include both elementary and middle schools. During the last 2 decades, the National Science Foundation has provided summer and academic year inservice training to over half of all the science teachers in the country to ensure understanding and personal investment in the new curriculum. It would be a contradiction of logic to suggest that so much expansion of the science curriculum could be responsible for student performance decline.

Several significant observations resulted from an analysis of the NAEP data. Tables 7.1 and 7.2 contain all of the decline and improvement findings along with some interpretations that have considered the variables in the context of the science curriculum.

The chapter is organized to unfold NAEP's findings through an explanation of the procedural stages that were used in the analysis. First, data-based studies from the general literature are cited. Then, highlights of the National Science Teachers Association (NSTA) analyses of the NAEP findings are described. Next, the average results of the three NAEP assessments are examined. This serves as a structuring device for the subsequent examination of the two groups of items showing the most positive and most negative changes. Finally, conclusions are drawn from the findings of the in-depth analysis.

REVIEW OF THE LITERATURE

The majority of the literature on test score declines consists of expository material with a limited number of research interpretations and virtually no experiments. This review will discuss only the data-based literature. Although this will limit the number of articles cited, the quality and clarity of the arguments should be enhanced, resulting in a more efficient documentation of the current state of the field.

Improvement and decline in science achievement crosses all educational

TABLE 7.1
Possible Explanations for Observed Upper Hinge Score Improvements

Age (Modal Grade Level)		
9 (4th)	13 (8th)	17 (11th)
1. Unusual number of biology items, mostly from heredity and sex education (35% of items), Table 7.14. This is interpreted as being associated with high student interest and, hence, improved performance.	1. Many more visuals than no visuals among the test items (62% to 38%), Table 7.13. This is interpreted to suggest that the visuals provided greater clarity for the items.	No factors identified
2. Large proportion of test items in the "common experience" category, Table 7.17. This is interpreted to suggest simpler content and, hence, better scores.	2. High number of applications, apply the rule, and principle/functional relationships, Tables 7.19, 7.20, and 7.21. This is interpreted to illustrate a good curricular match with current science curricula.	
3. High proportion of combined SES factors HIMT+PHS+FABC, Table 7.22. This is interpreted to suggest that students from families with high interest in education and rich extra-school experience will probably have more science exposure and, hence, better scores.		

levels and fields through graduate school. Upper elementary and secondary school declines have been reported by Copperman (1979), Welch (1980), the NSTA (1973 and 1976), and the NAEP (April, 1979). At the undergraduate and graduate levels, science test score declines have been reported on the College Board Admissions Test (CBAT) and the Graduate Record Examination, Advanced Test (GRE). Exceptions to the decline reports do exist; however, they are limited to only two areas of improvement: early elementary tests (Copperman, 1979) and the CBAT and GRE biology tests (ETS, 1975a). Other than that, general performance in science among U.S. students at all levels appears to be on a steady decline.

One of the most extensive data sources on science achievement is the NAEP. Three national testings have been conducted—the first during the 1969–1970 school year, the second during the 1972–1973 school year, and the third during the 1976–1977 school year. Each test population included students from three age levels: 9, 13, and 17. During the 3 testing years, 543

TABLE 7.2
Possible Explanations for Observed Lower Hinge Score Declines

Age (Modal Grade Level)		
9 (4th)	13 (8th)	17 (11th)
1. Unusually high number of science items requiring identification of formal laws, Table 7.20. This is interpreted to be associated with low student interest, and low relationship to the current science curriculum.	1. High grade level equivalent on readability measure, Tables 7.25 and 7.26 among released items. This is interpreted to suggest that students may not be able to answer the questions correctly because they cannot read the questions.	No factors identified
2. Large percentage of items above grade level (72%), Table 7.17. This is interpreted to suggest that the readability of the items makes them beyond the students' comprehension.		
3. Twice as many technical flaws in the AGL items, Table 7.18. This is interpreted to suggest that the flaws increased the student's confusion, and therefore made those items more difficult.		
4. High number of knowledge, Level 1 performance classes or multiple discriminations, Tables 7.19, 7.20 and 7.21. This is interpreted to suggest that factual recall is not consistent with the current science curriculum's emphasis on process and application.		
5. High grade level equivalent on readability measure, Tables 7.25 and 7.26. This is interpreted to suggest that students may not be able to answer correctly because they cannot read the questions.		

different cognitive questions were asked of approximately 72,000 students, with about 2500 students answering each question. Some test items were administered to two or three of the age levels because a limited number of test items were used in 2 or 3 of the testing areas. In order to manage such an extensive data base, a strategy was designed to search for patterns of improvements and declines as well as for possible variables that might explain the observed change patterns.

The NSTA (1973) published two study team reports commenting on the NAEP findings. The first report, concerned with the 1969–1970 findings, concluded with four judgments:

1. Females performed more poorly than males on all science subject areas sampled and at all age levels sampled.
2. Seventeen-year-olds in school performed better than young adults of approximately the same age out of secondary school in 21 out of 36 common test items.
3. Rural, black, southeastern, inner city, and female students performed less well than students in other geographic, ethnic, socio-economic, sexual, or racial groups sampled. (This judgment included the assumption that "the objectives assessed by the National Assessment of Educational Progress [were] appropriate to the general student population found in the public schools, and not to an elite group [p. 39]."
4. Seventeen-year-olds and young adults did poorly on the test items assessing knowledge in the area of sex education.

Each of the four judgments proffered by the NSTA have raised issues for future analysis and research. There is a need for more careful attitudinal investigations to discover why females consistently perform lower than males. The higher level of in-school young adult performance over out-of-school young adults suggests the need for studies on science retention in nonschool environments and for a demographic analysis of the characteristics of out-of-school versus in-school young adults of the same age. The third judgment raises the question of whether the objectives, and the test items used to sample the objectives, correspond with the public school science curriculum at the elementary and secondary levels. An assessment of the context of the sex education curriculum is suggested by the fourth judgment. Concerns raised by the first, second, and fourth judgments are beyond the scope of this analysis, but the concern raised by the third is part of the analysis designed for this chapter.

The second NSTA study (1976) committee commented on the 1972–1973 NAEP findings and compared the findings of the first and second administrations. The committee concluded that there was "a real decline in the test scores [p. 8]." The committee raised the question of whether the test items were actually measuring one's ability to *apply* fundamental scientific principles and facts or whether they were merely assessing recall of memorized information. A reasonable extrapolation of NSTA's concern would lead to a discussion of levels of intellectual complexity (Gagné, 1970, Walbesser, 1978); the question would then become whether or not the items are assessing chaining or functional relationships, each of which represents roughly opposite poles of intellectual demand.

A careful examination of the NAEP test score data was conducted with the presupposition that patterns observed within these data would yield a specific set of dimensions for a more in-depth analysis. Any analysis of the NAEP data was dependent upon the variables already present.

Before describing the components of the analysis strategy decided upon, a glossary of the variable definitions used is provided.

1. Age (A): Results are reported for 9-, 13-, and 17-year-olds.
2. Grade level for each age (GL): Results are reported for 9-year-olds in either grade 3 or 4, 13-year-olds in either grade 7 or 8, and 17-year-olds in grade 10, 11, or 12.
3. Levels of intellectual complexity (LIC): An adapted set of categories from Gagné's *Conditions of Learning* are used including chaining, multiple discrimination, concepts, multiple attributes, functional relationships, and problem-solving.
4. Level of parent education (PE): Results are reported in three categories: parents who did not graduate from high school (NGHS), at least one parent graduated from high school (GHS), and at least one parent who has had some post-high school education (PHS).
5. Performance classes (PC): There are nine distinct categories of human performance—naming, identifying, stating a rule, distinguishing, ordering, constructing, demonstrating, describing, and applying a rule (Walbesser, 1968).
6. Race (RA): Results are reported for blacks (B) and whites (W).
7. Region (RE): Results are reported for four regions: Northeast (NE), Southeast (SE), Central (C), and West (W).
8. Science Areas (SA): All science topics are related to physical and life sciences covered at the elementary, middle, and senior high school levels, as well as special elementary science topics named by:
   (S-APA) Science - A Process Approach
   (SCIS) Science Curriculum Improvement Study
   (ESCP) Earth Science Curriculum Project
   (BSCS) Biological Sciences Curriculum Study
   (CBA) Chemical Bond Approach
   (PSSC) Physical Science Study Committee
9. Sex (S): Results are reported for males (M) and females (F).
10. Size of community (SC): Results reported for the following categories: Big city (BC)—schools within city limits of cities having a 1970 census population over 200,000: Fringes Around Big Cities

(FABC)—schools within metropolitan areas served by cities that have a 1970 census population over 200,000 but outside the city limits; Medium City (MCTY)—schools in cities having a population between 25,000 and 200,000 but not classified in the "fringes around big city" category; and Smaller Places (SMPL)—schools in communities having population less than 25,000 but not classified in the "fringes around big city" category.

11. Taxonomic level (TL): Bloom *et al.* (1956), *Taxonomy of Educational Objectives: Cognitive Domain* categories: Knowledge, comprehension, application, and the composite category, analysis–synthesis–evaluation.

12. Technical Flaws (TF): The following technical flaws were identified:
    There is ambiguity in the stem or in the responses.
    The correct response is the longest response.
    The number of distractors departs from the most frequent number of responses commonly seen.
    There is a lack of parallelism in language or in form.
    There is an incomplete stem or another error within the item.
    There is a problematic visual condition: error with the illustration, graph, or graphic.
    A universal quantifier is present in the stem or in the responses.
    There is the possibility that two responses are correct.
    The responses "all of the above" or "none of the above" are used.
    The item is not related to a cognitive assessment of science.
    The correct answer is not necessarily present among the responses.

13. Test administration ages and years (TAAY): The months and years during which each age group was tested:
    Age 9 was tested January–February 1970, 1973, and 1977; age 13 was tested October–December 1969, 1972, and 1976; and age 17 was tested March–May 1969, 1973, and 1977.

14. Test item visual (TIV): Any illustration, graph, or other graphic which is part of the stem or the responses.

15. Type of community (TC): Results are reported for the following categories:
    Advantaged-Urban (HIMT): schools in or around cities with a population greater than 200,000 where a high proportion of the residents are in professional or managerial positions.
    Disadvantaged-Urban (LOMT): schools in or around cities with a population greater than 200,000 where a high proportion of the residents are on welfare or are not regularly employed.
    Extreme-Rural (ER): Schools in areas with a population under 10,000 where most of the residents are farmers or farm workers.

**Findings Already Reported by NAEP and
Inferences Suggesting Directions for the In-Depth Analysis**

The analysis conducted on the NAEP findings was organized around the results obtained from 7 variables: grade level, parent education, race, geographic region, sex, size of community, and type of community. The 7 variables contain 19 differentiating categories including 1 for modal grade level, 3 for parent education, 2 for race, 4 for region, 2 for sex, 4 for size of community, and 3 for type of community. Average performance on the three tests was computed for each of the seven variables as well as for the nation as a whole.

Table 7.3 summarizes the grade level results on all items used from the 3 assessment years. All three age levels show means above the national mean computed from the average performance on the three administrations for all students. However, grade 4 (age 9) and grade 8 (age 13) show consistent trends downward over the three assessments. These findings do raise the question of whether or not the declines are a reflection of some curricular discrepancy between the instructional program and the test items. Perhaps some problems with the instrumentation (the tests or methods of measurement) offer another possible explanation.

Data on the parent education variable, shown in Table 7.4, indicate results below the national mean for students whose parents are not high school graduates (NGHS) and results above the national mean for students from families with at least one parent in the post-high school category (PHS). A consistent downward trend toward the national mean is present in the 9-year-old, PHS category. The above and below national mean find-

TABLE 7.3
Differences in Percentage Correct Between the Mean of the Nation and the Modal Grade Level on All Three Assessments for Ages 9, 13, and 17

Ages (Modal Grade Level)		
9 (4)	13 (8)	17 (11)
(+,↓)	(+,↓)	(+,θ)

KEY

+ Above National Mean   ↑ Consistent Increasing Trend
− Below National Mean   ↓ Consistent Decreasing Trend
0 At National Mean   θ No Consistent Trend

Source: National Assessment of Educational Progress, 1979(a), p. 25.

TABLE 7.4
Differences in Percentage Correct Between the Mean of the
Nation and Parent Education on All Three Assessments for
Ages 9, 13, and 17

Parent Education	Ages (Modal Grade Level)		
	9 (4)	13 (8)	17 (11)
NGHS	(−,θ)	(−,θ)	(−,θ)
GHS	(+,↑)	(−,θ)	(+,↓)
PHS	(+,↓)	(+,θ)	(+,θ)

KEY

+ Above National Mean   ↑ Consistent Increasing Trend
− Below National Mean   ↓ Consistent Decreasing Trend
0 At National Mean   θ No Consistent Trend

Source: National Assessment of Educational Progress, 1979(a),
p. 24.

ings suggest the need to consider the socio-economic variables present in
the data in more depth.

Table 7.5 reports the NAEP findings on the racial variable. Performance
of white students was consistently above the national mean, while the performance of black students was consistently below the national mean. No
3-year trends upward or downward were reported.

Data from the geographic regions are shown in Table 7.6. The Northeast

TABLE 7.5
Differences in Percentage Correct Between the Mean of the Nation
and Race on All Three Assessments for Ages 9, 13, and 17

Race	Ages (Modal Grade Level)		
	9 (4)	13 (8)	17 (11)
B	(−,θ)	(−,θ)	(−,θ)
W	(+,θ)	(+,θ)	(+,θ)

KEY

+ Above National Mean   ↑ Consistent Increasing Trend
− Below National Mean   ↓ Consistent Decreasing Trend
0 At National Mean   θ No Consistent Trend

Source: National Assessment of Educational Progress, 1979(a),
p. 24.

TABLE 7.6
Differences in Percentage Correct Between the Mean of the Nation
and Geographic Region on All Three Assessments for Ages 9,
13, and 17

Region	Ages (Modal Grade Level)		
	9 (4)	13 (8)	17 (11)
NE	(+,θ)	(+,θ)	(+,θ)
SE	(-,θ)	(-,θ)	(-,θ)
C	(+,↓)	(+,θ)	(+,↑)
W	(+,θ)	(-,↓)	(-,↓)

KEY

+ Above National Mean	↑ Consistent Increasing Trend
− Below National Mean	↓ Consistent Decreasing Trend
0 At National Mean	θ No Consistent Trend

Source: National Assessment of Educational Progress, 1979(a),
p. 24.

and Central regions were above the national mean for all three age levels,
and the Southeast was below the national mean for all three age levels. The
13- and 17-year-olds from the West showed results below the national mean
and a consistent 3-year trend downward. This might raise the question of
whether the position effects with respect to the national mean are another
manifestation of the state of the science curriculum or a manifestation of
some other effect (such as economic investment in science education)
within the schools of the regions. The need for socioeconomic analysis was
again reinforced.

Table 7.7 shows males continuing to obtain performance results above
the national mean for all three age levels, while females continue to obtain
performance results below the mean. Findings for 17-year-olds do show
consistent downward trends toward the national mean for males and upward trends toward the national mean for females. (Perhaps the interest in
science as a career is growing among females and declining among males.
This might be a combined curricular and societal outcome.)

The size of community variable revealed several patterns of interest in
Table 7.8. Students from fringes of big cities (suburbs) and middle size
cities perform above the national mean for all three age levels. Big city
students show 3-year means below the national mean, and the findings for
big city 13- and 17-year-olds are on a consistent trend downward. These
findings suggest the need for further analysis of the socioeconomic dimen-

TABLE 7.7
Differences in Percentage Correct Between the Mean of the Nation
and Sex on All Three Assessments for Ages, 9, 13, and 17

Sex	Ages (Modal Grade Level)		
	9 (4)	13 (8)	17 (11)
M	(+,θ)	(+,↑)	(+,↓)
F	(−,θ)	(−,θ)	(−,↑)

KEY

+ Above National Mean   ↑ Consistent Increasing Trend
− Below National Mean   ↓ Consistent Decreasing Trend
0 At National Mean   θ No Consistent Trend

Source: National Assessment of Educational Progress, 1979(a),
p. 24.

sion as well as the "declining effectiveness" question raised about big city
schools.

The seventh and final variable examined was type of community. Table
7.9 reports the results of this variable as compared to the national mean.
Performance results above the national mean were observed for students of

TABLE 7.8
Differences in Percentage Correct Between the Mean of the Nation
and Size of Community on All Three Assessments for Ages 9,
13, and 17

Size of Community	Age (Modal Grade Level)		
	9 (4)	13 (8)	17 (11)
BC	(−,θ)	(−,↓)	(−,↓)
FABC	(+,θ)	(+,θ)	(+,θ)
MCTY	(+,θ)	(+,↓)	(+,θ)
SMPL	(0,θ)	(+,θ)	(+,θ)

KEY

+ Above National Mean   ↑ Consistent Increasing Trend
− Below National Mean   ↓ Consistent Decreasing Trend
0 At National Mean   θ No Consistent Trend

Source: National Assessment of Educational Progress, 1979(a),
p. 25.

TABLE 7.9
Differences in Percentage Correct Between the Mean of the Nation
and Type of Community on All Three Assessments for Ages 9,
13, and 17

Type of Community	Age (Modal Grade Level)		
	9 (4)	13 (8)	17 (11)
ER	(-,↑)	(-,↑)	(-,↑)
LOMT	(-,↑)	(-,↓)	(-,↓)
HIMT	(+,θ)	(+,↑)	(+,↓)

KEY

+ Above National Mean   ↑ Consistent Increasing Trend
− Below National Mean   ↓ Consistent Decreasing Trend
0 At National Mean   θ No Consistent Trend

Source: National Assessment of Educational Progress, 1979(a),
p. 25.

all ages in the urban-advantaged category. The urban disadvantaged and
the rural student performance are all below the national mean. Identical
with the big city trends, the urban disadvantaged students show consistent
trends downward for 13- and 17-year-old groups. The rural students show
consistent upward trends for 9-, 13-, and 17-year-olds. (Again, need for additional socioeconomic analysis was revealed.)

Table 7.10 combines the information from Tables 7.3–7.9 into three
clusters: (a) position from the national mean; (b) consistent trends over the
three administrations: and (c) the combined position and trend patterns.
The position data summary shows an almost even split between variable
components above and below the national mean. The consistent trend data
reveal that the majority of variable components (approximately two-thirds)
show no consistent pattern over the three administrations of science assessment. About one-sixth of all variable components show a consistent pattern upward over the three administrations and all three age levels. These
changes obviously deserve special attention along with the one-sixth of all
of the variable components showing a consistent downward pattern.

The combined patterns mirror the 9-, 13-, and 17-year-olds for both
positive upwards (+,↑) and negative upward (−, ↑) trends. The negative
downward (−, ↓) trends experience greater frequency with increasing age.
Causes for the upward and downward trends require the further attention
for both their curricular implications and possible causes.

An analysis of the NAEP data made one thing evident: that any change

TABLE 7.10
Frequency and Percentage of Results Above and Below the National
Mean for the Three Assessments on Each of the 19 Levels Comprising
the Seven Variables, and Frequency of Consistent Increasing and
Decreasing Trends for Ages 9, 13, and 17

Results	Age (Modal Grade Level)		
	9 (4)	13 (8)	17 (11)
Position from National Mean			
+	11 (58%)	10 (53%)	11 (58%)
−	7 (37%)	9 (47%)	8 (42%)
0	1 (5%)	0 −	0 −
Total	19(100%)	19(100%)	19(100%)
Consistent Trends			
↑	3 (16%)	3 (16%)	3 (16%)
↓	3 (16%)	5 (26%)	6 (31%)
0	13 (68%)	11 (58%)	10 (53%)
Total	19(100%)	19(100%)	19(100%)
Combined Position and Consistent Trends			
(+,↑)	1 (5%)	2 (11%)	1 (5%)
(+,↓)	3 (16%)	2 (11%)	3 (16%)
(+,θ)	7 (37%)	6 (32%)	7 (37%)
(−,↑)	2 (11%)	1 (5%)	2 (11%)
(−,↓)	0 −	3 (16%)	3 (16%)
(−,θ)	5 (26%)	5 (26%)	3 (16%)
(0,↑)	0 −	0 −	0 −
(0,↓)	0 −	0 −	0 −
(0,θ)	1 (5%)	0 −	0 −
Total	19(100%)	19(101%)*	19(101%)*

*Differs from 100% because of rounding.

in item results which may have occurred in two or three administrations of
the test should be the focus of attention. To effectively isolate those
changes, the usual strategy of analyzing only the most extreme change data
was adopted.

Under ordinary circumstances, when only one test administration is involved, the upper and lower 27% of the items would be analyzed. This
selection process affords the least possibility for overlap between the
groups as well as the least computational error in item difficulty and item
discrimination. But the NAEP data of interest did not reflect direct item

performance—only *change* scores were analyzed. In this case, the change scores were the differences observed on average results for each item used in at least two administrations. Since the upper and lower change score hinges were analogous to the high and low 27% for direct item results, it was decided that only those change score items which appeared in the upper and lower hinges of an age group would be analyzed. So, for an item to appear at or above the upper hinge, it had to come from the group of the most significant positive change items observed over the last two administrations of the NAEP tests. Similarly, a lower hinge item came from the group of the most significant negative change items observed over the last two administrations of the NAEP tests. This choice helped to reduce the likelihood that any isolated trend or effect would be related to measurement error.

Final support for the choice of hinge analysis came with the realization that NAEP had applied the "jackknife" procedure (Mosteller & Tukey, 1968) to their data, which, by the nature of its statistical process, provides fewer biased estimates for a population of scores that is subdivided into many different groups—in the NAEP case, 19 differentiating categories within 7 variables.

The next step was to choose the dimensions of the analysis. They were: (*a*) the number of years the item was used; (*b*) the socioeconomic conditions affecting performance; (*c*) the readability of the items; (*d*) the appropriateness of the content to the age–grade level; (*e*) the technical flaws in the item; (*f*) the age–grade levels at which the item was used; (*g*) the taxonomic level of each item; (*h*) the performance class level of each item; and (*i*) the level of intellectual complexity of each item.

Findings and Interpretations

The upper and lower hinges were chosen as the focus for an in-depth analysis of possible causes of the science assessment score changes. Identifying the hinges also provided a framework for examining the most significant positive and negative changes in the data. The extremes analysis strategy maximizes the likelihood of locating any helpful variables and minimizes the possibility of including confounding variables or conditions.

Past examinations of these data failed to identify a single explanation for the changes. Therefore, this study's assumption is that no single variable or combination of variables is likely to exist to explain all of the changes—improvements or declines—across all or even one age level.

Using the NAEP data analysis presented in the *Technical Summary Report* (April, 1979), the upper and lower hinge items were closely examined for possible patterns to explain the significant positive and negative

TABLE 7.11
Distribution of Test Items by Age and Number of Years Used
Within Each Hinge Level

Hinge/Years	Age (Modal Grade Level)		
	9 (4)	13 (8)	17 (11)
Upper			
2	58%	90%	71%
3	42%	10%	29%
Lower			
2	50%	71%	58%
3	50%	29%	42%

changes. Some attention was also given to any variables which could help distinguish between the upper and lower hinge items (See Table 7.11).

More items with a history of being used during two as opposed to three test administrations are reported for both the upper and lower hinges. There are fewer 3-year than 2-year items. Both of these findings are predictable and do not suggest any curricular or instructional impact.

Some of the items in the NAEP assessments have been released; others remain unreleased. The question of whether the upper and lower hinges are affected by the released or unreleased status of the items is examined in Table 7.12.

Findings in two of the six groupings do not follow the expected pattern of more released than unreleased test items. The lower hinge 9-year-olds and the upper hinge 13-year-olds show the discrepant patterns, and may help to account for any special effects noted in subsequent analysis of additional variables.

TABLE 7.12
Distribution of Released and Unreleased Test Items by Age
Within Each Hinge Level

Hinge/Status	Age (Modal Grade Level)		
	9 (4)	13 (8)	17 (11)
Upper			
Released	69%	43%	71%
Unreleased	31%	57%	29%
Lower			
Released	33%	86%	67%
Unreleased	67%	14%	33%

Visuals play a clarifying role when paired with oral or written tests. Acknowledging this well-known relationship, the obvious conjecture is that upper and lower hinge placement is a function of the presence or absence of visuals to support the written portion of the test items. Table 7.13 presents these data.

In most cells, there are fewer test items *with* visuals than without. This seems to be true for both the upper and lower hinges of all grade levels. (The one interesting exception is the upper hinge 13-year-olds, a cell noted earlier. In this cell, there is a reversal, and 62% of the items contain visuals.)

The upper and lower hinge items were also analyzed by content area, for an appropriate match of specific content with each grade level assessed could have a bearing on the overall performance on each item. One might expect, therefore, that most general science items would appear in the 9-year-old tests; that most earth science items would appear in the 13-year-old tests; and that most biology, physics, and chemistry items would appear in the 17-year-old tests. By and large, Table 7.14 confirms these expectations with two notable exceptions: the upper hinge 9-year-old items were 35% biology-related, and the lower hinge 9-year-old items were 22% earth science-related.

The upper hinge 9-year-old biology test items were almost all on heredity or sex education. These are high interest content areas which appeal directly to the individual's knowledge about himself and his origins. This might very well explain some of the position change results observed in this cell.

The lower hinge 9-year-old earth science test items are almost all about functional relationships, which require little more than direct memorization and are, consequently, of low interest to students. The content of these

TABLE 7.13
Distribution of Test Items with Visuals by Age
Within Age Level

Hinge/Visual	Age (Modal Grade Level)		
	9 (4)	13 (8)	17 (11)
Upper			
Visual	32%	62%	24%
No Visual	68%	38%	76%
Lower			
Visual	6%	50%	8%
No Visual	94%	50%	92%

TABLE 7.14
Distribution of Test Items By Age and Subject Area With
Hinge Level

Hinge/Subject	Age (Modal Grade Level)		
	9 (4)	13 (8)	17 (11)
Upper			
General Science	50%	67%	24%
Earth Science	5%	14%	12%
Biology	35%	14%	40%
Chemistry	–	–	6%
Physics	5%	5%	18%
N/A	5%	–	–
Lower			
General Science	61%	54%	17%
Earth Science	22%	21%	8%
Biology	11%	11%	33%
Chemistry	–	–	25%
Physics	6%	–	17%
N/A	–	14%	–

items is often taught without laboratory or other direct student involvement. All of this suggests that the presence of these particular test items is one factor in the declining results observed for the lower hinge test items in the 9-year-old test group.

All test items are subject to both measurement errors and technical errors which are introduced during the construction of an item. Standard errors have been reported separately by the NAEP staff documents but no special insight into why the particular items were in the upper or lower hinges was provided (NAEP, 1979a, pp. 12, 13, 15).

An in-depth technical error analysis was conducted using the technical errors named in the definition section of this chapter. The reasoning was that more technical errors were likely to create confusion and, therefore, more likely to result in lower performance. Taken one step further, the conjecture was that there would be a higher frequency of technical errors in the lower hinge than in the upper hinge. The technical flaw data reported in Table 7.15 certainly do *not* support this conjecture. By examining the zero errors rows for both the upper and lower hinges, it is apparent that in two age categories, ages 9 and 17, just the reverse is true. There are more technical errors among the upper hinge items than the lower hinge items. Results for the age 13 group are about equal for both hinges (19 and 18% flaw-free). Even though it does not seem very sensible to believe that there would be better performance with poorly constructed items, "technical flaws" did not seem to be a variable that explained the observed difference.

TABLE 7.15
Distribution of Test Items with Technical Flaws Within Hinge
Groups and By Age

Hinge/Flaws	Age (Modal Grade Level)		
	9 (4)	13 (8)	17 (11)
Upper			
0	21%	19%	12%
1	47%	65%	29%
2	26%	12%	47%
$\geq$ 3	5%	4%	12%
Lower			
0	39%	18%	25%
1	28%	57%	42%
2	22%	18%	33%
$\geq$ 3	11%	7%	−

Even when the analysis separated released and unreleased items to examine technical flaws (returning to the notion that the unreleased items were somehow different from the released items within the two hinges) and explored the question of whether "unreleased" items might house a substantial portion of the errors, the patterns shown in Table 7.16 do not support this conjecture. The distribution follows the same pattern as the pattern of released to unreleased test items shown in Table 7.12. The most reasonable conclusion, therefore, is that technical flaws combined with released or unreleased status of items do not appear to explain any of the upper or lower hinge results.

Are the test items appropriate to the modal grade level of the students be-

TABLE 7.16
Percentage of Test Items With At Least One Technical Flaw for
Released and Unreleased Test Items Within Hinge Levels and By Age

Hinge/Released	Age (Modal Grade Level)		
	9 (4)	13 (8)	17 (11)
Upper			
Released	47%	31%	65%
Unreleased	32%	35%	24%
(No Flaws)	(21%)	(34%)	(11%)
Lower			
Released	21%	68%	50%
Unreleased	37%	14%	25%
(No Flaws)	(42%)	(18%)	(25%)

ing assessed? That is, are the 9-year-old items appropriate to the fourth grade, the 13-year-old items appropriate to the eighth grade, and the 17-year-old items appropriate to the eleventh grade? With the assistance of a subject matter expert and a sampling of textbooks used during each of the three test administrations, each item in the upper and lower hinge was classified as follows: appropriate to grade level (GL), above grade level (AGL), or part of most individuals' out-of-school common experiences (CE). The results of this classification are summarized in Table 7.17.

Three observations appear to be of some special note. First, the vast majority of upper and lower hinge test items are on grade level for the 13- and 17-year-old audiences. Second, a very large part of the age 9 items in the lower hinge are above grade level. This alone may well account for the declining scores among the 9-year-olds. Third, the largest proportion of "common experience" test items are present in the upper hinge 9-year-old category. This may account for the improving change scores for the upper hinge 9-year-old items.

Is there a relationship between appropriateness to grade level, presence of technical flaws, and upper or lower hinge placement with the age categories? Table 7.18 explores these possible relationships. Note that legitimate upper and lower hinges are only possible within an age category because the three age groups represent independent populations. This table provides reinforcement for the explanation that the proportion of the total set of items and technical characteristics of the "above grade level" items are directly linked to the observed 9-year-old test score decline. (This observation might also explain the general 9-year-old decline illustrated earlier in Table 7.3.) The AGL items are likely to be the most difficult, and are likely

TABLE 7.17
Distribution of Test Items by Age and Appropriateness Within Hinge Level

Hinge/Appropriateness	Age (Modal Grade Level)		
	9 (4)	13 (8)	17 (11)
Upper			
CE	26%	10%	–
GL	32%	71%	89%
AGL	42%	19%	11%
Lower			
CE	11%	–	8%
GL	17%	82%	75%
AGL	72%	18%	17%

TABLE 7.18
Distribution of Test Items by Appropriateness and by Presence of
At Least One Technical Flaw

Hinge/Flaw/Appropriateness	Age (Modal Grade Level)		
	9 (4)	13 (8)	17 (11)
Upper			
0			
CE	11%	5%	–
GL	11%	14%	12%
AGL			
Subtotal	22%	19%	12%
=1			
CE	16%	5%	–
GL	26%	57%	76%
AGL	37%	19%	12%
Subtotal	79%	81%	88%
TOTAL	101%[a]	100%	100%
Lower			
0			
CE	–	–	8%
GL	11%	11%	8%
AGL	28%	7%	8%
Subtotal	39%	18%	24%
=1			
CE	11%	–	–
GL	6%	71%	67%
AGL	44%	11%	8%
Subtotal	61%	82%	75%
TOTAL	100%	100%	99%[a]

[a]Differs from 100% due to rounding.

to be made more confusing by technical flaws. This heightens the difficulty
further. For 9-year-olds in the lower hinge with at least one flaw, two-
thirds of the items are in the AGL (above grade level) category. Therefore,
the proposed explanation for at least the 9-year-old decline becomes a bit
more plausible.

The analysis of the content in terms of appropriate grade level placement was followed by an examination of the objectives. Two general classification systems were used to assess them: taxonomic level and intellectual complexity level. Bloom's *Taxonomy of the Cognitive Domain* (1956) is the classical framework of the first. NAEP reduced Bloom's system to four levels: knowledge, comprehension, application, and a combined category—analysis/synthesis/evaluation.

Levels of intellectual complexity are an adaptation of Robert Gagné's categories described in *The Conditions of Learning* (1970). According to this system, learning experiences can be classified into one of five categories: chains, multiple discriminations, concepts, principles, and problem-solving, each of which is intended to represent an order—from simple to complex—of learning difficulty.

Both Bloom's and Gagne's taxonomies address general categories of intellectual demand, but neither relates directly to *specific* behaviors named by performance objectives. Walbesser's (1968) nine performance classes constitute a sufficient set of categories to organize all stated objectives. They are (*a*) name; (*b*) identify; (*c*) state a rule; (*d*) order; (*e*) distinguish; (*f*) construct; (*g*) demonstrate; (*h*) describe; and (*i*) apply the rule. Walbesser's performance classes are also accompanied by three learning time levels. Contrary to conventional operational design, effective learning and testing cannot be accomplished without attention to both specific behavior and time required to perform. Walbesser's system allows the classroom teacher, the curriculum designer, and the testing specialist to satisfy both concerns through the application of one classification system.

Three separate tables of findings are presented for taxonomic level, performance class, and levels of intellectual complexity. As shown in Table 7.19, the taxonomic patterns across age levels are very much the same for both the upper and lower hinges. In some respects, this seems contrary to the curricular expectation that lower levels of the taxonomy should gradually decrease in frequency as age and grade level increase. Higher levels of the taxonomy might be expected to exhibit the reverse of the lower level pattern, moving from less to more instances of higher levels with increasing age and grade level. This observed balance is possibly best explained by assuming a deliberate attempt on the part of the test makers to retain such balance across age levels.

Comprehension items are in the majority in all six cells with knowledge and application items sharing the second most frequent occurrence. There are two exceptions to this pattern. First, a much larger percentage of knowledge items is reported in the lower hinge for 9-year-olds than in any other cell. Second, the largest percentage of application items and the smallest percentage of knowledge items are reported in the upper hinge for 13-year-olds. Before commenting on the possible meaning of each of these

TABLE 7.19
Distribution of Test Item Objectives by Taxonomic Level[a] Within
Hinge Levels for Ages 9, 13, and 17

Hinge/Taxonomic Level	Age (Modal Grade Level)		
	9 (4)	13 (8)	17 (11)
Upper			
Knowledge	19%	10%	12%
Comprehension	63%	57%	71%
Application	12%	33%	18%
Analysis/Synthesis/ Evaluation	6%	–	–
Lower			
Knowledge	33%	16%	17%
Comprehension	50%	53%	50%
Application	17%	26%	25%
Analysis/Synthesis/ Evaluation	–	5%	8%

[a]For detailed definitions see Bloom (ed.), Taxonomy of
Educational Objectives: The Cognitive Domain, 1956.

pattern exceptions, the information about "performance class" and "levels
of intellectual complexity" will be discussed first.

Performance class distribution is shown in Table 7.20. The percentages
within each level of the performance show two patterns which may help explain the following two observations. The lower hinge 9-year-old cell is
overwhelmed by Level 1, "identify objectives" (83%). The upper hinge
13-year-old results contain the smallest overall percentage of Level 1 performance class objectives (38%) and the largest percentage of Level 3 performance class objectives (52%).

The third facet to the objectives analysis is summarized in Table 7.21.
The "levels of intellectual complexity" pattern shows the same general
balance across hinge levels and age categories, much like what we have seen
in Tables 7.19 and 7.20. In considering the findings in all three tables, two
hunches emerge: one about the lower hinge 9-year-old results, the other
about the upper hinge 13-year-old results. The large percentage of chains
and multiple discriminations present in the lower hinge 9-year-old cell
coupled with the large number of identify objectives (Table 7.20) and the
large number of knowledge objectives (Table 7.19) all suggest a rather
high density of memorizable and factual information. Therefore, the
declining test scores for this particular cell may be a reflection of the shifting focus of elementary school science *toward* more process and applica-

TABLE 7.20
Distribution of Test Item Objectives by Performance Class[a] Within Hinge Levels for Ages 9, 13, and 17

Hinge/Performance Class		Age (Modal Grade Level)		
		9 (4)	13 (8)	17 (11)
Upper				
Name		6%	–	9%
Identify	Level 1	69%	33%	55%
State a Rule		–	5%	–
Order	Level 2	–	–	–
Distinguish		6%	10%	–
Construct		–	–	–
Demonstrate	Level 3	–	–	–
Describe		–	14%	9%
Apply a Rule		19%	38%	27%
Lower				
Name		–	–	–
Identify	Level 1	83%	59%	44%
State a Rule		–	–	–
Order	Level 2	–	–	–
Distinguish		17%	18%	19%
Construct		–	6%	–
Demonstrate	Level 3	–	–	–
Describe		–	6%	25%
Apply a Rule		–	12%	13%

[a]For detailed definitions see Henry H. Walbesser, Competency-Based Curriculum, 1978.

tion objectives, and *away from* memorized factual information as influenced by the elementary school science curriculum projects of the 1960s such as S-APA[1] and SCIS.[2]

Following the same line of argument, the substantial number of items at the "principle-functional relationship" level in the upper hinge 13-year-old cell, the high percentage of "apply the rule" items reported in Table 7.20, and the high percentage of "application" items reported in Table 7.19 combine to suggest the explanation that the significant improvement changes are a result of testing performances most closely related to the students' curricular experiences.

[1]American Association for the Advancement of Science, Commission on Science Education. *Science - A Process Approach.* 1966.

[2]University of California at Berkeley. *Science Curriculum Improvement Study.* 1964.

TABLE 7.21

Distribution of Test Item Objectives by Levels of Intellectual Complexity[a] Within Hinge Levels for Ages 9, 13, and 17

Hinge/Level of Intellectual Complexity	Age (Modal Grade Level)		
	9 (4)	13 (8)	17 (11)
Upper			
Chain	13%	10%	–
Multiple Discrimination	25%	10%	31%
Concept	13%	14%	–
Principle-Multiple Attributes	13%	5%	25%
Principle-Functional Relationship	37%	57%	38%
Problem Solving	–	5%	6%
Lower			
Chain	22%	11%	9%
Multiple Discrimination	39%	42%	27%
Concept	–	–	–
Principle-Multiple Attributes	–	21%	9%
Principle-Functional Relationship	39%	26%	55%
Problem Solving	–	–	–

[a]For detailed definitions see Robert M. Gagne, The Conditions of Learning, and Henry H. Walbesser, Competency-Based Curriculum.

Socioeconomic status (SES) has a history of use as a variable that might explain test score results as well as direct performance assessment differences observed among groups of students. Information gathered by the NAEP allowed for the construction of a pseudo-SES variable by combining parent employment, parent education, and urban–suburban location. High SES was defined as HIMT (advantaged-urban) combined with PHS (post-high school parent education) and FABC (fringes around big city). Low SES was defined as LOMT (disadvantaged-urban) combined with NGHS (parents who did not graduate from high school) and BC (big city). A search was made of the upper and lower hinge items. An item was classified as a high SES provided that the percentage count reported for HIMT was the highest for all categories with the racial category excluded since this would confound the definition chosen for SES. Similarly, an item was identified as low SES provided that LOMT was the smallest percentage count for all categories. Whenever HIMT and the post-high school dimension parent education variable (PHS) or the suburban dimension of the size of community variable (FABC) were the two highest percentages, this condition was noted. A similar tally was made for the combinations of the three low SES dimensions: LOMT, NGHS, and BC. Table 7.22 summarizes the SES analysis.

TABLE 7.22
Percentage of All Test Items by Age and Socio-economic Status
Within Hinge Level

Hinge/SES	Age (Modal Grade Level)		
	9 (4)	13 (8)	17 (11)
Upper			
High SES			
HIMT	26%	4%	12%
HIMT + (PHS or FABC)	11%	12%	35%
HIMT + PHS + FABC	53%	23%	24%
Combined	90%	39%	71%
Low SES			
LOMT	26%	8%	6%
LOMT + (NGHS or BC)	37%	19%	12%
LOMT + NGHS + BC	37%	19%	47%
Combined	100%	46%	65%
Lower			
High SES			
HIMT	22%	36%	17%
HIMT + (PHS or FABC)	17%	14%	42%
HIMT + PHS + FABC	17%	29%	17%
Combined	56%	79%	76%
Low SES			
LOMT	33%	25%	17%
LOMT + (NGHS or BC)	11%	29%	25%
LOMT + NGHS + BC	22%	32%	42%
Combined	66%	86%	84%

The one striking observation about the SES analysis was how similar the high and low SES patterns were within each of the six hinge-age cells. The SES does not seem to help explain either the upper or lower hinge results with one possible exception. The large proportion of HIMT + PHS + FABC entries does suggest the possibility that the improvement change scores which comprise the upper hinge for the 9-year-olds may be a result of this particular student population. The argument might be that students from such an SES category are more likely to come from families who place a high value on education and school performance. Those students are also more likely to have been exposed to more out-of-school experiences related to general science information. Another interpretation of the SES data is derived from the balance between the "combined high

SES" data and the "combined low SES" data which are present for all ages and hinge levels. The entire pattern might be explained by the observation that this is a homogeneous grouping effect. Bright students from both high and low SES categories are contained in the upper hinge and slower students from both SES categories are producing the lower hinge data. No test of this conjecture was possible with the NAEP data at hand, but could well be explored by future research.

One of the response alternatives available for each item on the NAEP science test was the response "I Don't Know." Table 7.23 summarizes these data for the hinge and age categories. The large percentage of change observed for the lower hinge across all age groups suggests a growing test-wiseness (or a reluctance to guess) which may be part of the increased use of, and experience with, accountability measures throughout the country.

If the student did not answer the question, NAEP recorded an entry in the "no response" column. The upper and lower hinge analysis of the "no response" data is presented in Table 7.24. The upper hinge items show a decrease in this category across all age groups. This may be evidence of an improving match between the content included on the test and the content as covered through classroom instruction. At least the finding suggests that the content is recognizable. The most dramatic evidence of change is the substantial increase in the use of "no response" among the lower hinge 13-year-old data. Further examination, however, revealed that this increase is a result of students not responding to: (a) one item with six separate parts; (b) an item where the correct answer may not even be present among the choices; and (c) an item not directly related to science achievement.

TABLE 7.23
Distribution of Test Items by Significant Change in the "I Don't Know" Category for Age and Within Hinge Levels

Hinge/"I Don't Know"	Age (Modal Grade Level)		
	9 (4)	13 (8)	17 (11)
Upper			
Increase	16%	14%	18%
Decrease	26%	10%	29%
No Change	58%	76%	53%
Lower			
Increase	78%	50%	42%
Decrease	–	11%	8%
No Change	22%	39%	50%

TABLE 7.24
Distribution of Test Items by Significant Change in the "No Response"
Category for Age and Within Hinge Levels

Hinge/"No Response"	Age (Modal Grade Level)		
	9 (4)	13 (8)	17 (11)
Upper			
Increase	5%	–	6%
Decrease	16%	10%	18%
No Change	79%	90%	76%
Lower			
Increase	–	32%	–
Decrease	6%	–	–
No Change	94%	68%	100%

Hence, the finding provided no real additional insight into the test score decline of 13-year-olds.

Can the students read the test items? An assessment of the readability of the test items in both the upper and lower hinges was made. Released and unreleased test items were examined separately. The Fry measure of readability was used, since it is more sensitive to the lower and upper end of the K–12 grade levels than is the FOG readability measure. Table 7.25 summarizes the findings on test item readability for the released and unreleased items.

Concerning appropriate readability, the desired grade level equivalent suggested for each age level is based upon the instructional research supporting the relationship of one grade level reduction for each 4 grade levels (e.g., for students in the eighth grade, the appropriate reading level of the

TABLE 7.25
Measures of Readability in Grade Level Equivalents of Released
and Unreleased Test Items for Ages 9, 13, and 17

Test Items	Age (Modal Grade Level)		
	9 (4)	13 (8)	17 (11)
Released	5th	8th	7th
Unreleased	7th	5th	6th
Desired	3rd	6th	9th

test items should be sixth grade; for students in the twelfth grade, the appropriate reading level would be ninth grade). This relationship holds only for instructional or assessment material intended for independent reading and/or processing. Close examination of the test items clearly establishes appropriate reading levels for the 17-year-old items and the unreleased 13-year-old items. The 9-year-old items, however, show reading levels in excess of both the desired reading level and the modal grade level. High reading levels, particularly among the unreleased items, is another plausible contributor to the 9-year-old test score decline. The eighth grade level for the released test items for the 13-year-olds likewise points to readability as a contributor to 13-year-old test score decline.

Table 7.26 summarizes the readability data for the upper and lower hinges and confirms the observations made from Table 7.25. The hinge data do isolate the lower hinge 13-year-old cell with an eighth grade readability level, and reinforce the plausibility of the explanation that readability is a contributor to the decline observed in the lower hinge 13-year-old data.

Summary of Findings and Interpretations

The in-depth analysis was conducted around the upper and lower hinge items for 9-, 13-, and 17-year-olds. These data included both released and unreleased test items. Variables examined as potential explainers of the change score improvements and declines included years of use, presence of visuals, socioeconomic status, content topic, appropriateness to grade level, readability, technical flaws, significant "I Don't Know" answers, significant "no response," and common item across age, taxonomic level, performance class, and levels of intellectual complexity. The problem of test score decline was broken down into six cells, three related to the upper hinge for ages 9, 13, and 17, and three related to the lower hinge for the

TABLE 7.26
Measures of Readability in Grade Level Equivalents of Upper and Lower Hinges for Ages 9, 13, and 17

Hinge	Age (Modal Grade Level)		
	9 (4)	13 (8)	17 (11)
Upper	5th	6th	9th
Lower	5th	8th	7th
Desired	3rd	6th	9th

same age groups. The singular composite problem was translated into six independent problems. One principal advantage of this reductionist strategy is to reduce the level of universality required of any potential explaining variable.

The findings appear to offer substantial insight into three of the six conditions: upper hinge 9-year-olds, upper hinge 13-year-olds, and lower hinge 9-year-olds. One possible lead is supplied for the lower hinge 13-year-old condition. No explanations emerged for either condition relating to 17-year-olds. Tables 7.1 and 7.2 (introduced at the beginning of the chapter) identify those findings. A reexamination of the clarifying tables alluded to in each finding description might prove helpful.

CONCLUSIONS AND RECOMMENDATIONS

No single variable or set of variables was uncovered as the general cause of the test score decline across all three tested age levels. The strategy of separating the problem into smaller parts was a useful one, however, because it facilitated drawing conclusions about 9- and 13-year-olds.

9-year-olds

Conclusion 1: Change score improvement among 9-year-olds is explained by three variables, two of which are directly related to the content, and the third, indirectly related. The first two speak to high interest content, and content settings which can be drawn from personal experience. The third variable identifies the high SES factor through which more cultural experience related to science content is likely to occur.

The conclusion elicits commentary related to both curriculum and social conditions which can influence effective instruction. It is interesting to note that the section of the test showing the greatest positive change contained items drawing upon the students' high interest or familiar experiences. This suggests the need for renewed considerations from both the classroom instructor and the test maker. Laboratory experiences designed to enable students to master scientific principles might be more effectively modified by altering the choice of setting and equipment to include more directly relevant experiences for the student. Using the same logic, the test designer might give more scrutiny to the context of the stems which are constructed to measure the student's comprehension.

There is some question, though, as to whether common experience items should be present in any national assessment, since there are few, if any, experiences which can be relied upon as universal ones for any age group.

The SES variable relates to curriculum from the affective domain; it does remind the educator that parental attitudes toward education do influence student achievement.

Conclusion 2:   Change score decline among 9-year-olds is directly related to five variables: (a) low student interest, low relationship to current curriculum; (b) content beyond·grade level of instruction; (c) large percentage of items with technical flaws; (d) emphasis on assessing lower levels of intellectual skills; and (e) readability which is above grade level.

The test score decline among the 9-year-old change scores seems to be a product of an ill-advised assortment of items with shortcomings in content, intellectual level, and appropriateness. Unfortunately, the emphasis on factual recall is inconsistent with the elementary school curriculum reform of the past 2 decades, as discussed earlier. Added to this is a significant number of items with technical flaws and high reading levels, both of which are likely to contribute to increased item ambiguity and confusion for the student. The recommendation that most obviously follows from this conclusion is that item content be aligned more closely with actual curricula, and that more quality control be maintained over the items in order to reduce the flaws and control the readability. Lastly, more performances should be sampled at the application level or upper range of intellectual complexities.

13-year-olds

Conclusion 1:   Change score improvement among 13-year-olds is related to the presence of two conditions: (a) a high proportion of visuals per set of test items, and (b) a high percentage of items requiring application of fundamental relationships.

The impact of the visual medium as an effective clarifier is well-known. Test items which combine the visual and the symbolic (words) factors are likely to obtain a more accurate measurement of the presence or absence of a given behavior. Perhaps more deliberate attention should be given to the inclusion of visuals in test items. The high frequency of fundamental relationship applications is consistent with the contemporary science curriculum of the middle school, and such content sampling should logically continue.

Conclusion 2:   Change score decline among 13-year-olds is related in part to a high reading level among the test items.

If the students cannot read the questions, it is not likely that their performance will accurately reflect the behaviors they possess. Readability checks should be conducted as a routine part of test item construction prior to any field testing or general use.

No insight was obtained from this analysis for the 17-year-old declines or improvements and, therefore, no conclusions about this age group are offered.

Certain general observations result from the analysis. The question of what constitutes the science curriculum for any particular age level is central to an examination of the objectives assessed and the content used to sample those objectives. No general information of this kind was available. NAEP could provide great assistance by gathering this important data base through requesting the following from participating schools: (a) the titles of the science textbooks being used: (b) the percentage which accurately reflects the students' attendance rates in the school; (c) the science curriculum enrollment figures by course; and (d) whether or not, according to the teacher's appraisal, the test's objectives and item content are appropriate for the age level being tested.

Speaking more directly to the objectives, the NAEP statements *identified* as objectives are not descriptions of observable performances and, by virtue of their more general form, increase confusion about the actual performance domains being sampled. Stating clear objectives that are observable performances would assist both the test designer and those interested in analyzing the science test data, not to mention classroom instructors committed to competency-based curricula.

The test items examined here seem to sample very heavily from the lower levels of the taxonomy, performance class, and intellectual complexity classifications for all ages sampled. A redistribution of the test items to include all of the levels would provide greater insight into the nature of the score declines.

NAEP has made a major contribution to the improvement of public education in the United States. Continued efforts at national assessment should be supported. Clearly, documentation of score decline and improvement has the positive benefit of both alerting educators to needs for change and acting as a catalyst to bring about improvements in the educational process.

ACKNOWLEDGMENTS

The authors wish to acknowledge the following individuals for contributing their expertise in the construction of this chapter: content verification, Sally Ann Cooper, secondary science instructor, Howard County Public School System and the Howard County Department of Education professional library, Columbia, Maryland; data clarification, David Wright, Coordinator of Scoring and Analysis Procedures, National Assessment of Educational Progress,

Denver, Colorado; item verification, Vern Achterman, Assessment Services Coordinator, National Assessment of Educational Progress, Denver, Colorado; and data acquisition, University of Maryland, Baltimore County campus, Baltimore, Maryland.

REFERENCES

Bloom, B.S. (Ed). *Taxonomy of educational objectives: The classification of educational goals.* Handbook 1. *Cognitive Domain.* New York: McKay, 1956.

Copperman, P. The achievement decline of the 1970s. *Phi Delta Kappan, 60:*736–739, June, 1979.

Educational Testing Service. *Graduate record examinations supplemental interpretive data.* Princeton, New Jersey, 1975a.

Educational Testing Service, *Guide to the use of the graduate record examinations, 1975–1977.* Princeton, New Jersey, 1975b.

Gagné, R. *The conditions of learning* (2nd ed.). New York: Holt, Rinehart and Winston, 1970.

Heikkenen, H. Science test score decline in Maryland. Paper presented to University of Maryland faculty, College Park, Maryland, December 3, 1975.

Miller, R.G., Jr. A trustworthy jackknife. *Annals of mathematical statistics,* No. 35, pp. 1574–1705, 1964.

Mosteller, F. and Tukey, J.W. Data analysis including statistics. In Lindzey, G. and Aronson, E.(Eds.), *Handbook of social psychology* (2nd ed.). Reading, Massachusetts: Addison-Wesley, 1968.

National Assessment of Educational Progress. *Three national assessments of science: changes in achievement, 1969–1977.* Denver, Colorado, June, 1978a.

National Assessment of Educational Progress. *Science achievement in the schools.* Denver, Colorado, December, 1978b.

National Assessment of Educational Progress. *Three assessments of science 1969–1977: Technical summary.* Denver, Colorado, April, 1979a.

National Assessment of Educational Progress. *Attitudes towards science.* Denver, Colorado, October, 1979b.

National Science Teachers Association. National assessment findings in science 1969–1970: What do they mean? *The Science Teacher, 40*(6), 33–40, September, 1973.

National Science Teachers Association. *An assessment of NAEP: Are the science declines real?* Washington, D.C., January, 1976.

Walbesser, H. H. *Competency-based curriculum.* Beltsville, Maryland: Walbesser and Associates, 1978.

Walbesser, H. H. *An evaluation model.* Washington, D.C.: American Association for the Advancement of Science. 1968.

Welch, W. W. A possible explanation for declining test scores, or learning less science but enjoying it more. *School Science and Mathematics, 80*: 22–28, January, 1980.

Chapter **8**

Trends in Educational Standards in Great Britain and Ireland

Thomas Kellaghan
George F. Madaus

In this chapter, we shall examine trends in scholastic attainment in Great Britain and Ireland. A perusal of comments on standards of performance in schools and of attempts to deal with the perceived inadequacies of the educational system over the past hundred years indicates that concern with standards is not a recent phenomenon. However, it is really only since World War II that objective tests have been used to monitor standards, and the use of such tests has begun, at least, in all five areas which are the concern of this chapter: England, Scotland, Wales, Northern Ireland, and the Republic of Ireland.

The main focus of this analysis will be on the use of standardized tests in surveys designed to assess national standards. Such studies were concerned with students up to 15 years of age. For evidence on the performance of older students, we shall consider achievements on public examinations that are geared to secondary school curricula. Not a great deal of information on national standards can be gleaned from public examinations. Indeed, any information that can be obtained is incidental to the purpose for which public examinations are taken at all, which is to assess the performance of individual students, and the relevant research has been carried out in the context of comparability of standards rather than to assess trends in attainment. Nevertheless, public examination performance is worth considering for a number of reasons. As indicated, it provides evidence on older students which is not available from other sources. Further, the type of in-

The Rise and Fall of National Test Scores
Copyright © 1982 by Academic Press, Inc.
All rights of reproduction in any form reserved.
ISBN 0-12-068580-9

formation it provides differs from that which is typically sought in studies of trends in attainment, that is, norm-referenced standardized test information. We shall attend to the advantages and disadvantages of public examinations, as well as those of standardized tests, for the purpose of monitoring national standards of attainment.

ASSESSMENTS OF STANDARDS WITHOUT THE USE OF TESTS

Prior to using standardized tests for the examination of national trends in attainment, more informal methods were used, and indeed such methods continue to be used to some extent today. Historically, they are of interest in that they indicate a concern with standards that stretches back well into the last century.

These methods took a number of forms; all were basically informal and impressionistic in nature. This was supported in the testimony given to government-appointed commissions set up to investigate various aspects of education, annual reports of schools' inspectors, and opinions of concerned laymen and politicians. We cannot review all such information here; however, a few examples may provide something of its flavor.

We can go back to 1868 to record negative comments on standards of attainment in elementary schools. In that year, a Royal Commission of Inquiry into Primary Education in Ireland was undertaken by the Earl of Powis. Having received testimony and examined evidence, the commission concluded that "the progress of the children in the national (elementary) schools of Ireland is very much less than it ought to be [Ireland, 1870]." The commission, being particularly concerned about the poor standard of English, recommended as a remedy the adoption of a scheme in which teachers' salaries would be dependent in part on the results of annual examinations in reading, spelling, writing, and arithmetic. It was felt that the scheme, known as "payment-by-results" would be "an effective inducement to more work on the part of teachers [Ireland, 1870]."

Payment-by-results is of interest, not just for its early attention to the concept of accountability, but also because it set a pattern common in British education in that examinations dictated to a large extent what was taught in schools and how it was taught.

It was not surprising, for a system that was introduced in the context of accountability, that a number of approaches were taken to assess the system itself. First, the percentage of pupils examined and passed was considered. Sir Patrick Keenan (1881), in an address to the National Association for the Promotion of Social Sciences, reviewed the scheme when it had been in operation for 10 years. He compared the educational returns of 1870, before the system of results was in operation, with those of 1880, when the system was in full operation, and was able to report good results. In the 10-year period, the percentage of students passing in reading had increased from 70.5 to 91.4, in writing from 57.7 to 93.8, and in arithmetic from 54.4 to 74.8. He further noted that the 1880 figures were better than those obtained in England with the exception of arithmetic, for which figures were similar in both countries.

A second method of assessing the payment-by-results scheme was to examine literacy figures in dicennial censuses. The education authorities were pleased to note that illiteracy had declined from 33% of the population in 1871 to 14% in 1901. However, Coolahan (1977) points out that the decline in illiteracy was largely the same in the 30 years preceding the payment-by-results era.

As a third method of assessing the payment-by-results scheme, schools' inspectors judged the quality of the reading they observed in schools. Their conclusions were less optimistic than those based on examination passes or literacy figures. They pointed out that pupils could pass the reading test without actually being able to read. In 1895, one inspector reported that reading is "generally indistinct in articulation, the words are so falsely grouped and proper emphasis and pauses so much neglected, that it has become often an unintelligible jumble" (Coolahan, 1977, p. 15).

At the turn of the century, following the abandonment of the payment by results system (which understandably had always been strongly opposed by teachers), school inspectors first reported an improvement in reading standards. This was so even though the "revised programme," which accompanied the demise of payment by results, removed from the curriculum a heavy emphasis on the 3 Rs, and emphasized a more child-centered approach and introduced a wide range of subjects, including manual ones (Coolahan, 1977). By the 1920s, however, criticisms of standards of attainment were again being made. In the Irish Free State (later the Republic of Ireland), where a new, more subject-centered primary school curriculum was introduced following its independence, a Commission on Technical Education noted in 1927 that constant reference was made by representatives of educational associations and by committees of technical institutions "to the difficulties experienced in the technical school with students whose primary education was defective [*Irish School Weekly*, 1928, p. 268]." In Northern Ireland also, where the "revised programme" continued to operate following partition, it was reported to the Lynn Commission that pupils appeared unduly retarded (Great Britain: Northern

Ireland, 1923–1934). Thus, whether the school curriculum was subject-centered, as in the Irish Free State, or pupil-centered, as in Northern Ireland, its products were similarly criticized.

Such criticisms of scholastic standards were frequently echoed in parliamentary debates. On November 12, 1925, for example, a deputy in the Irish parliament complained that primary education had deteriorated and "that if it continues to deteriorate at the present rate, we must emerge in a short time as almost a nation of illiterates [Ireland: Dail Eireann, 1925, p. 269]." In reply, the Minister for Education indicated his belief that school standards were as high as they ever had been. He focused on the difficulty in comparing one period with another, pointing out that "every one of us knows that as people advance in life they are always discovering all the excellent things that existed 20, 30, 40 years ago, sometimes rightly, but sometimes it is a sort of imagination that develops with advancing years [Ireland: Dail Eireann, 1925, p. 364]."

The difficulties in making comparisons over time were further developed by a later Minister for Education in 1941. Faced once again with criticisms of standards, he pointed to the lack of empirical evidence on the matter, indicating that something more than our own recollections was needed. Further, he recognized that even if information on performance over time on comparable tests were available, changes in social and educational circumstances would have to be taken into account in interpreting trends (Ireland: Dail Eireann, 1941, p. 1219).

The inadequacies of the kind of evidence we have been considering scarcely need much comment. There is really very little that we can learn about trends in standards from the comments contained in reports of commissions, schools' inspectors or parliamentary debates. Many of the commentators, with the exception of school inspectors, would have had little day-to-day contact with schools. Their opinions most likely would have been based on limited and perhaps untypical evidence about particular schools or particular sets of students. They are also unlikely to have taken into account factors that might affect the general level of attainment such as changes in the general population structure and in the school population in particular.

Census figures were of some value, even though they were based on personal reports, when rates of literacy, however defined, were changing considerably. Today, the vast majority of people profess some level of literacy, and census data no longer discriminate sufficiently among degrees of literacy to provide any useful information on trends over time.

The point made by the minister last cited that there was a need for more objective evidence in assessing trends over time in standards of attainment was well taken, and it was the recognition of this need that led to the use of objective tests to monitor attainment. In the next section we shall consider information derived from the use of such tests.

ATTAINMENTS ON STANDARDIZED TESTS

Most of the information available on attainment is on the subject of reading, though in recent years, through the work of the Assessment of Performance Unit (APU) of the Department of Education and Science in Britain, the range of subjects covered in the monitoring of standards has been broadened.

Reading Surveys in England

In England, surveys of reading go back to 1948, and attempts have even been made to relate post-war to pre-war standards. Indeed, one of the purposes in conducting the 1948 survey was to obtain data that would be relevant to suggestions being made at the time that "backwardness" and "illiteracy" had increased during the war (Great Britain: Ministry of Education, 1957).

Between 1948 and 1977, a series of surveys to measure reading standards was carried out. Two tests were used in these surveys: the Watts–Vernon Test and the National Survey Form Six (NS6). Both are tests of "silent reading comprehension," made up of sentence-completion type items. Each item consists of a sentence with one word missing; the pupil's task is to read the sentence and select from among five words the most appropriate one to complete and make sense of the sentence. There are 35 items in the Watts–Vernon test and 60 in the NS6. In both tests, items become progressively more difficult. There is a time limit of 10 minutes for the Watts–Vernon test and one of 20 minutes for the NS6. Only the Watts–Vernon test, which was constructed in 1947, was used in the earlier surveys. Since 1955, the NS6 test, which was constructed in 1954, has been used either in conjunction with the Watts–Vernon or on its own.

The surveys focused on two groups: pupils in their last compulsory year of schooling (age 15) and pupils in their last year of primary schooling (age 11). To assess the attainments of 15-year-olds, the Watts–Vernon test was administered in 1948, 1952, 1956, 1961, and 1971 and the NS6 in 1955, 1960, and 1971. In the case of 11-year-olds, the Watts–Vernon was administered in 1948, 1952, 1956, 1964, and 1970 and the NS6 in 1955, 1960, 1970, and 1976–1977. In 1976–1977, 9-year-olds were tested for the first time on the NS6.

Descriptions of the findings are to be found in a number of sources. (For the earlier surveys, see Great Britain: Ministry of Education, 1950, 1957,

1960; for the later ones, see Great Britain: Department of Education and Science, 1963, 1966, 1973, 1978; Peaker, 1967; Start & Wells, 1972.) The results from the testings are presented in raw scores in Table 8.1. Two points should be borne in mind when reading the table. First, the 1948 samples were judgment samples, not random ones; however, the standard error was calculated on the assumption that the sample was random. Second, the 1961 sample of 15-year-olds was confined to secondary modern and comprehensive schools; the mean score in the table is an estimate for the populations represented in the other surveys on the assumption that schools other than secondary modern ones achieved the same increase in standards. The standard error, however, was calculated from the actual survey data (see Start & Wells, 1972).

The trends in the table are fairly clear. In general, for the 11-year-olds, level of performance on the Watts–Vernon test increases from 1948 to 1964 and drops slightly in 1970. No further data are available on this test for this age group. Similarly, on the NS6, mean performance rises from 1955 to 1960 and stays at about the same level in 1970. However, a further testing in 1976–1977 indicates resumed improvement of the standard of performance. The 15-year-olds also show an improvement from 1948 to 1961 on

TABLE 8.1
Mean Performance (and standard error) on Reading Tests in England, 1948 to 1977.

Year	11 year olds				15 year olds			
	Watts–Vernon		NS6		Watts–Vernon		NS6	
	M	SE	M	SE	M	SE	M	SE
1948	11.59	0.59			20.79	0.37		
1952	12.42	0.30			21.25	0.20		
1955			28.71	0.55			42.18	0.64
1956	13.30	0.32			21.71	0.26		
1960			29.48	0.52			44.57	0.73
1961					24.10	0.14		
1964	15.00	0.21						
1970	14.19	0.38	29.48	0.92	23.46	0.26		
1971							44.65	0.83
1976/77			31.13	0.33				

the Watts–Vernon and from 1955 to 1960 on the NS6. Further testing in 1971 indicates no further improvement in performance. More recent information comparable to that for the 11-year-olds is not available for the 15-year-olds.

In addition to raw scores, reading age scores (a popular concept in British tests) were also used to describe trends. In the early surveys, the average score appropriate for children of a specific age was decided on a simple, observational basis. For example, if 15 were the average score for 11-year-olds, then a child with a score of 15 was regarded as having a reading age of 11 years (Great Britain: Ministry of Education, 1957). More recently, an age allowance per month was estimated for groups by calculating the gradient of the regression line when scores are plotted against age in months. The gradient indicates the average improvement made each month by pupils in that age group, while its reciprocal represents the number of months of reading age that correspond to 1 point of score (Start & Wells, 1972).

Statements about improvement in reading standards were presented in terms of reading age. For example, it was concluded that 11-year-olds tested in 1964 "reached on average the standard of pupils 17 months older in 1948 [Start & Wells, 1972, p. 22]." Similarly, for 15-year-olds, an improvement of nearly 20 months of reading age was recorded in the period 1948 to 1961 (Start & Wells, 1972). Further, the improvement was greater in the first half than in the second half of the decade. Statements such as these, couched in terms of reading age, seem readily understandable and have an obvious appeal to teachers and the general population.

A final attempt in describing trends was to introduce an element of criterion referencing. Thus, it was argued that the early questions in the Watts–Vernon test "could be read and answered by an intelligent child in the infants' school," while the later questions needed "the same vocabulary and understanding as the leading article in a good newspaper [Great Britain, Department of Education and Science, 1966, p. 2]."

A more ambitious effort at criterion referencing was the attempt to relate the test scores to levels of literacy. In comparing the results of the early surveys, pupils were classified on the basis of their reading ages, which were calculated on the basis of the average reading score for an age group in 1938. Six categories were used: superior, average +, average −, backward, semi-literate (a reading age of between 7 and 9 years), and illiterate (a reading age less than 7 years). The last two categories were combined for the 11-year-olds (Great Britain: Ministry of Education, 1950). From 1948 to 1956, comparisons were made of the proportions of pupils who fit into each category, and it was found that the proportion of 11-year-olds classifiable as semi-literate or illiterate had dropped from 5% in 1948 to 1% in 1956. For the 15-year-olds, the drop for the same categories over the same period was from 6% to 4%. At the other end of the scale, the proportion of "superior" 11-year-olds increased from 9 to 17%; however, the proportion of "superior" 15-year-olds remained at 9% (Great Britain: Ministry of Education, 1957). The proportion of poor readers classified in this way continued to decrease in later surveys (Start & Wells, 1972).

These data confirm the analyses based on raw scores. In particular, they indicate that when improvement occurs it takes place across the board and that weaker students benefit as much, if not more, than more able ones.

While the rise in standards until about 1960 was generally well received, the argument was made that it indicated nothing more than a return to standards that had existed before the war and that, not surprisingly, had been adversely affected during the war years (Burt, n.d.). Unfortunately, no ready comparison could be made between pre-war and post-war data to substantiate this claim. Only information from local samples of children was available for 1938, and considerable variation was found by region. Also, pooling regional data might not produce a very satisfactory national sample. Further, the test used in 1948 had not previously been used. To attempt some kind of comparison, it would be necessary to re-administer the pre-war tests to a sample of pupils in 1948. This was in fact done. The rather tentative conclusion arrived at was that "the proportions of backward readers in English and Welsh primary and secondary modern schools are larger than they were before the war [Great Britain, Ministry of Education, 1950, p. 48]." Thus, some support was provided for the view that the rise in standards through the 1940s and 1950s was a result of the return to more normal living and educational conditions after the war.

During the 1960s and 1970s, there was considerable criticism of standards in school, some of which seemed to be supported in the findings of the 1970 surveys. The criticism was expressed most strongly in the so-called "Black Papers" on education and was focused on the role of "progressive" teaching methods, which were perceived as responsible for a general lowering of scholastic standards. In correspondence received by the Bullock committee, there was a majority view that standards of literacy had declined, or, at any rate, were at a standstill (Great Britain: Department of Education & Science, 1973). Perhaps the findings of the most recent survey (1976–1977) have in some way allayed fears on reading standards in primary schools.

Reading Surveys in Other Countries

Information on standards of reading in other areas is less extensive than it is for England. Until 1955, a sample of Welsh children was included in

England's trend surveys. That data for the 1948 and 1952 surveys relate to English children only; however, in the 1955 survey, there was no distinction between English and Welsh pupils.

In 1956, England's survey was paralleled in Wales. Subsequently, reading surveys were carried out in Wales in 1960 and 1971. The results of these three surveys—1956, 1960 and 1971—are reported in Horton (1973). The tests used (the Watts–Vernon and the NS6) and the age levels (11- and 15-year-olds) surveyed were the same as England's. The Watts–Vernon was used with both age groups in 1956 and in 1971; the NS6 was used with both groups in 1960 and in 1971. A Welsh language test was also administered in 1971.

For the 11-year-olds, a significant rise in standards on the Watts–Vernon was found between 1956 and 1971. The difference between the 1960 and 1971 scores on the NS6 (a slight decrease) was not found to be significant. In the case of the 15-year-olds, significant improvements were recorded on the Watts–Vernon between 1956 and 1971 and on the NS6 between 1960 and 1971.

When the Welsh results are compared with England's, a significant difference in favor of English children on the Watts–Vernon is found for both 11- and 15-year-olds in 1956 (11-year-olds in Wales: 12.01; in England: 13.30; 15-year-olds in Wales: 20.76; in England: 21.71). On the 1960 NS6 test, English 15-year-olds (Wales: 42.29; England: 44.57), but not 11-year-olds (Wales: 28.58; England: 29.60), were still scoring significantly higher than Welsh pupils of comparable age. A comparison of the Welsh 1971 data and the English 1970 data indicates that, while English children scored slightly higher than Welsh children, differences between the two groups were not significant. (On the Watts–Vernon, 11-year-old English children scored 14.19 and Welsh children 13.86; 15-year-old English children scored 23.46 and Welsh children 23.04. On the NS6, 11-year-old English children scored 29.38, Welsh children 28.58; 15-year-old English children scored 44.96, Welsh children 44.78).

Periodic national surveys of attainment have not been carried out in Scotland; thus, information about standards is very limited. In one survey, reading standards of pupils in Primary 7 class (about age 11), as measured by the Edinburgh Reading Test Stage 3, were found not to have changed significantly between 1972 and 1975 (Maxwell, 1977).

In Northern Ireland, the first national reading attainment survey of 11-and 15-year-olds (using the NS6 test) was carried out in 1972 (Wilson, 1973). No data on trends over time are available. A comparison of the scores of Northern Ireland children with those of English children indicates that at 11 years of age the difference between the groups in mean score on the NS6 is not statistically significant (Northern Ireland: 29.14; England:

29.85), while at 15 years it significantly favors English pupils (Northern Ireland: 42.15; England: 44.96) (Wilson, 1973).

In the Republic of Ireland in 1964 the first survey of reading comprehension was carried out (Kelly & McGee, 1967). Unfortunately, it was confined to the Dublin area. The survey used the NS6 test of reading comprehension which had been used in English surveys and focused on 11-year-olds. The survey was repeated in 1969 (McDonagh, 1973), 1974 (Travers, 1976; McGee, 1977), and 1979 (McGee, 1980). A significant improvement was not recorded between 1964 and 1969; however, since then there has been a steady improvement. Over the 15-year period (1964–1979), the mean score has risen from 19.95 to 28.21 (McGee, 1980). Compared to the performance of English children, the 1964 score of Irish children was extremely low (mean score of English 11-year-olds in 1960 was 29.48). The 1979 Irish score is still lower than the 1976–77 English score (31.13); however, Irish children have moved considerably toward closing the gap noted in earlier surveys.

Attainments in Areas Other Than Reading

The assessment of attainment trends in areas other than reading has begun only recently and so no data on trends are yet available. The Assessment of Performance Unit (APU) was set up in Great Britain's Department of Education and Science in 1975 to monitor, over time, three main aspects of pupils' development—language, mathematical, and scientific—in England and Wales. Language assessment began in 1979, mathematical assessment in 1978, and scientific assessment in 1980. The initiation of the project reflects concern about educational standards and accountability. As in the reading surveys, attention is focused on 11- and 15-year-olds.

Results of the first mathematical survey are available (Great Britain: Department of Education and Science, 1980). Obviously, no comparative data were available with which to compare these results. The results provide a detailed description of 11-year-old pupils' knowledge of concepts, skills, and applications of mathematics. Practical as well as written tests were included. Finally, an attitude questionnaire sought evidence on pupils' "liking" of mathematics and their perceptions of the "difficulty" and "utility" of the subject.

The information which will be derived from the APU surveys will provide data on trends in attainment of a far more detailed nature than the kind of information that was available from the reading surveys. Thus, it is hoped, it will be of more interest and practical use than was the reading survey data to teachers, parents, employers, and educational administrators.

TRENDS IN STANDARDS IN PUBLIC EXAMINATIONS

Performance on public examinations geared to school curricula seem particularly appropriate for assessing the output of schools (Madaus *et al,* 1979). In Britain and Ireland, such examinations form an important part of the educational system at the secondary level. The examinations are set by independent examining boards and are taken by students at about 16 and 18 years of age.

In this section we will consider some evidence relating to trends in standards on the Certificate of Secondary Education (CSE) and on the General Certificate of Education (GCE), which, for the most part, are taken by students in England and Wales. The CSE, taken at 16 years of age, is intended for students in the lower ability range. There are two levels of GCE: "ordinary," taken at age 16 and intended for students in the higher ability range; and "advanced," taken at age 18 and intended for students in the highest range of ability. Large numbers sit for the examinations: In 1975, over 620,000 students were following CSE and GCE ordinary level courses in maintained schools (Bush, 1979).

Since the examinations are regarded as important by students, teachers, and the general public, there is strong pressure on schools to teach the subject matter and skills which are examined. Thus, evidence on changes in examination standards should directly reflect the work of schools. Further, since there are examinations in a wide range of subjects (over 60), it should be possible to monitor quite specific areas of achievement.

It is generally assumed that grades in a subject on public examinations signify the same level of attainment from year to year (Bush, 1979). While attempts have been made to check this empirically, such attempts have been fraught with serious difficulties. For example, before considering comparability over years, one has to contend in any one year with comparability across subjects (see Bardell *et al.,* 1978; Forrest, 1971; Forrest & Smith, 1971) as well as comparability across boards of which there are 8 at the GCE level and 14 at the CSE level. A major problem in the assumption of equivalence, however, relates to differences in the reliability of examinations in different subject areas (Willmott & Nuttall, 1975). Comparability is a problem of considerable practical importance, for grades granted in the examinations are treated as equivalent by test users (e.g., universities and employers). It is not surprising that most research relating to comparability has been concerned with reliability and equivalence across subjects and boards rather than with equivalence across years.

If standards in fact are stable over years, then one possible index of trends in attainment would be the percentage of students who pass a given subject over a number of years. Pass rates for a number of commonly taken subjects at GCE ordinary level (English language, history, Latin, mathematics, physics, and biology) for the years 1970 to 1976 indicate considerable consistency in pass rate (approximately 60%) for most subjects (Bardell *et al.,* 1978). However, this information casts no light on the basic problem: does a pass in different years really represent equivalence in attainment from year to year or is the consistency primarily a function of a decision to keep the proportion of passes fairly constant? If the latter situation is the case, then maintainence of standards for a greater number of students could indicate a lowering of standard of attainment over time. We will return to this point later.

In one study of comparability of standards, this question was directly addressed since an attempt was made to take into account the ability level of candidates (Willmott, 1977). In addition to taking public examinations (CSE and GCE ordinary level), candidates also sat for a standardized test of general ability, and these scores were used as a reference or common standard against which examination performance each year could be assessed. Differences between years in grade were estimated after taking into account differences in the test ability of the examination candidates. The years 1968 and 1973 were chosen for comparison.

It was found that, with one exception, the quality of entry for all subjects, as assessed by the general ability test, had fallen between 1968 and 1973. Further, in all school subjects except two, grades were awarded more leniently in 1973 than in 1968. The exceptions were technical drawing in the CSE examination and French (taken by boys) in the GCE examination. On both tests, the standard for girls seemed to have fallen more than the standard for boys. In the CSE examinations, standards in art, French, biology, mathematics, and history seem to have altered the most, while in the GCE examinations, standards in chemistry, history, art, and English were the ones most affected.

Several reservations must be entered in considering the findings of this study. Analysis that is based on reference test scores assumes that "the ability of teachers to promote learning in their pupils (perhaps by inspiring and motivating their pupils) *is much the same in 1973 as it was in 1968* [Willmott, 1977, p. 81]." Criticisms of the study relate to sampling, particularly the adequacy of the GCE sample and the pooling of the grades of the individual boards (Bush, 1979). It should also be added that if the ability measured by the reference tests is not directly relevant to the learning of the subject matter examined in the public examinations, then the use of reference tests as a criterion measure may be misleading.

If, despite reservations, it is accepted that standards in public examinations have fallen, then what does that tell us about the output of schools? Obviously, at an individual level, employers, universities, and other users may be concerned that an examination grade earned by a student represents a somewhat lower level of attainment in 1973 (and perhaps more recently?) than it did in 1968. However, if we take into account the numbers of students passing the examinations (even at a slightly lower level), which have increased considerably during the time in question,[1] then it is clear that the "attainment output" of the educational system has increased considerably. A rough indication of that increase could be obtained by calculating the product of grade levels (converted to a numerical value) and the number of candidates achieving the grades for the two years in question.

A consideration of public examination performance does not provide us with a great deal of information on national standards of attainment, and it may be that the system is so complex that it will never be possible to obtain unambiguous and clear evidence from this source. However, as already indicated, the type of subject-specific, curriculum-sensitive test employed in public examinations provides a good "curriculum-test" match (Harnischfeger & Wiley, 1977), and this is likely to be more useful in assessing educational standards than the traditional type of norm-referenced standardized test which has often been used for this purpose. To provide adequate information on trends in attainment, however, it may be necessary to use the same kind of test as is used in public examinations in a more controlled way.

PROBLEMS IN MONITORING NATIONAL STANDARDS

The experience of monitoring national attainment trends over the past 30 years in Britain, even though the areas of concern have been limited, has sensitized investigators to the problems involved in the task. Indeed, many of the problems have been discussed in considerable detail in trend reports (e.g., Great Britain: Department of Education and Science, 1973; Start & Wells, 1972). Here, we may consider these problems briefly as they relate firstly to sampling of students and secondly to sampling of attainment.

Sampling of Students

It might appear that the ideal situation for trend assessments is one in which a representative sample of the same population took the same test at appointed intervals. As in any kind of research, there are problems in obtaining representative samples for national assessments. In the English

---

[1]The number of candidates taking at least one CSE subject examination rose from 207,045 in 1968 to 304,732 in 1972. During the same period, the number taking at least one GCE ordinary subject examination rose from 253,189 to 303,233 (Wilmott, 1971).

reading surveys, different sampling approaches were taken in different surveys. The 1948 sample was a judgment sample, the 1952 and 1956 samples were three-stage (area, school, pupil) probability samples, and the 1961 and 1964 samples were two-stage (school, pupil) probability samples. Since each method of sampling has its own implications for standard errors of means, different problems arise in comparing standards achieved in different surveys.

Changes also occurred over time in the populations being sampled. This problem arises partly from the complexity of the British educational system, in which different types of schools operate at both the elementary and secondary level. We saw, for example, that the 1961 survey of 15-year-olds was confined to students in secondary modern and comprehensive schools because the inquiry with which it was connected was interested in the less able students in the population. In the other surveys, pupils in private and special schools were not included. To the extent that pupils in such schools differ from the rest of the population, the figures obtained in surveys will not present a true picture of national norms. This, of course, will not affect comparisons over time for the population that is assessed if that population remains stable. However, such stability cannot be assumed. Indeed, we know that special educational provision increased over the life of the surveys. Thus, later samples would have included fewer slow learners than earlier ones.

Participation rate is always a problem in large-scale surveys, and the national surveys that we have considered did not escape this problem. For example, in the 1970 reading survey of 11-year-olds, only 73% of the schools selected agreed to participate. The response to the 1971 survey of 15-year-olds was even worse. The operation of this survey was complicated by a postal strike with the result that only 53% of schools agreed to participate. By contrast, in the 1976–1977 survey, all schools responded and the only loss of data (3 out of 540 schools) was due to errors in school administration. It is perhaps not without interest that the 1970 and 1971 surveys were conducted by a private research organization (The National Foundation for Educational Research in England and Wales), while the 1976–1977 survey was carried out by Her Majesty's Inspectors of Schools.

Problems also arose in the reading surveys because of differences in the ages of students surveyed. For example, in 1952 and 1956, each age group covered an age range of three months, centered on 15.0 and 11.0 years. In 1961, the average age of the senior students was below 15, while that of the junior students was higher than 11 years (Great Britain: Department of Education & Science 1966). To deal with this problem, age corrections were made on the basis of the regression line gradient that was obtained when scores were plotted against age in months. The appropriateness of such age corrections, particularly as ages deviate from the mean age, is open to question (Start & Wells, 1972).

In the case of examinations that operate beyond the age of compulsory schooling and for which students can exercise choice (whether or not to take an examination, which courses they will study, and which examinations they will opt for), problems with population changes assume particular importance. The magnitude of population changes can be seen when we consider that the number of GCE ordinary level candidates rose from 134,000 in 1951 to 950,000 in 1976 (Bardell *et al.*, 1978). If selective factors operate differentially for subjects in different years, difficulties in making comparisons become obvious, even if use is made of a reference test.

Sampling of Attainment

A test should provide an adequate sample of the skills and content that comprise the attainment which one wishes to assess (e.g., reading, mathematics, chemistry). The selection of items that will represent any specified universe of attainment, however, presents its own problems, and these have been recognized in the reading surveys that England carried out and in the movement toward more specific attainment tests as represented in the work of the Assessment of Performance Unit of the Department of Education and Science.

Since reading was the area that received the most attention in attainment studies, it is not surprising that the tests used were paid special attention. Both the Watts–Vernon and the NS6 tests were limited to one type of test item. The narrowness of the measure was recognized: At best, the tests could only be said to attempt to measure comprehension, and even in the area of comprehension, they did not measure inferential aspects of reading (Great Britain: Department of Education & Science, 1973). While attempts to specify the subskills of reading comprehension have been limited in their success (Davis, 1968; Hunt, 1957), the use of a variety of approaches has nonetheless been recommended by a number of commentators (Farr, 1969; Ward, 1956). The problem was avoided to some extent in the British surveys. For example, it was claimed that one of the tests used "implies a definition of reading ability that is in accordance with common sense [Great Britain: Department of Education & Science, 1966, p. 15]." Further, correlations between the type of test that was used and other reading tests (e.g., word recognition tests) were found to be high [about .80] (Great Britain: Ministry of Education, 1950).

This is not to say that aspects of the tests were not closely scrutinized for inadequacies. The content of items was examined for possible biases for different groups in the population. For example, it was suggested that the lower average performance of girls on the Watts–Vernon test was probably a function of the content of the test which used material that might be more interesting to boys than to girls.

A problem that arises when measures are used over a long period of time is that the measures may not adequately represent curricula and syllabi which change over time. The Bullock report noted that the problem was "one of attempting to assess the product of a variety of contemporary aims and methods with instruments constructed years ago [Great Britain: Department of Education & Science, 1973, p. 16]." It has been argued that words such as "haberdasher" and "mannequin" were more common in the 1940s and 1950s than in the 1970s; thus, pupils at the later date would be at a disadvantage by comparison with pupils who had taken the tests earlier (Start & Wells, 1972). It is unlikely that these standardized tests are particularly sensitive to changes in the content of school curricula, though they might be sensitive to changes in processes and attitudes associated with curricula as well as to students' experience with particular kinds of examinations and tests.

A further problem in using the same test over a long period of time is that if there is a marked improvement in performance, there may not be room on the test to register the improvement. Thus, if the test has not sufficient "head room" ("ceiling," in American parlance [Eds.]), it will not allow the whole distribution to move upward in score. The consequences of using such a scale are that the mean score will be depressed and the distribution will not be normal. In this situation, measures of standard error will not be the same when the distribution is normal (Great Britain: Department of Education & Science, 1973). It has been demonstrated that the reading tests used in British surveys did not satisfactorily discriminate among the more able 15-year-olds in later surveys (Great Britain: Department of Education & Science, 1973), while in the 1976–1977 survey, some of the ablest 11-year-old students achieved nearly maximum scores (Great Britain: Department of Education & Science, 1978). These data indicate that the more recent surveys may have underestimated increases in reading standards in the population.

Another factor which may have inhibited performance in more recent surveys is Britain's relative lack of familiarity with standardized tests. From 1970 onward, it seems reasonable to assume that pupils have had considerably less experience with objective, standardized tests than pupils in earlier surveys when a battery of standardized tests played an important role in selection for secondary school (Start & Wells, 1972).

Familiarity with tests may also have been a factor in the results of the surveys in Ireland, though in this case, it would have worked in the opposite direction. Given the most recent performance of Irish pupils in the reading attainment survey, which phenomenally increased over 10 years, it is most unlikely that the earlier results were a true reflection of their reading achievement. Changes in primary education in Ireland since 1970, including the introduction of a new curriculum, have meant that pupils have a greater familiarity with standardized tests than in the past and, more importantly, with the kinds of items which appear in such tests. For example, the kinds of exercises that occur in many reading books, which include questions quite similar to some of the items in standardized tests, only appeared in Irish reading books during the last decade. Further, as long as a public examination was employed at the end of primary schooling (a practice discontinued in 1967), the attitudes and practices of pupils were no doubt affected by that examination. By contrast with standardized tests, that examination asked essay questions and placed a premium on students taking care in reading the examination questions and on their deliberation in answering.

Problems with standardized tests are not confined to students' degree of familiarity with such tests. A more basic concern is to ensure that the test adequately samples skills and contents that constitute attainment in the area under consideration. There is already some evidence that norm-referenced standardized tests are relatively insensitive to school-specific achievement and, instead, tend to measure general traits that are strongly influenced by home background variables (Madaus et al., 1980). Certainly, British and Irish public examinations seem more sensitive to what goes on in schools. At a prima facie level, tests of English literature, history, chemistry, and biology would seem more appropriate to assess the attainments of students than general tests of reading and mathematics. The former type of test taps content, skills, and processes which are closely related to school instruction. Not surprisingly, when performance on such school-specific examinations is empirically related to school characteristics, the relationship is apparent (Brimer et al., 1978; Madaus et al., 1979).

If one is interested in national trends irrespective of educational conditions, then whether the test used is curriculum-sensitive or not might not be very important. But when educational policy is in question, a nation will most likely be interested in the antecedents of attainment and the possibility of remedial action. In this situation, some attention to the role of the school seems inevitable.

The APU's movement in Britain toward the assessment of a large range of specific attainments indicates a realization of the inadequacies of norm-referenced standardized tests for assessing national standards. Because of the country's tradition of public examinations, we may expect to see considerable development in this area in the coming years. Not only should this result in a more detailed picture of students' attainments (which can be monitored over time) than was available from norm-referenced tests, but it should also provide information that will form a basis for policy decisions about remedial action if necessary.

CONCLUSION

In the light of the problems considered, that is, student sampling and ensuring an adequate representation of school attainment in examinations, there are obvious difficulties in drawing any firm conclusions about trends in attainment standards on the basis of the surveys that have been carried out over the past 30 years. However, if we bear in mind that any conclusions we might draw must be subject to the reservations imposed by a consideration of such problems, a number of general statements may be made.

As far as reading standards are concerned, elementary school children (11-year-olds) may be said to have improved since World War II. Some of the improvement immediately following the war may only indicate a return to prewar standards. During the 1960s, there was a slight reversal in the general trend toward improvement, but the reasons for this are not clear. One possible explanation is that the reversal reflects a change in the population of the country, particularly in the number of immigrant children from non-British cultural backgrounds who came to Britain after the war. According to this explanation, the resumed rise in reading standards in the 1970s would suggest that educational problems associated with immigration had decreased.

The evidence for a rise in the educational standards of secondary school students is less obvious than it is for elementary school students. The reading attainments of 15-year-olds, as was the case for 11-year-olds, showed an improvement up to 1960; however, no further improvement was recorded after that date. This failure to show further improvement may have been the result of using a test which did not permit discrimination among the best students. Thus, limitations of the test may have precluded a demonstration of an actual improvement in achievement standards after 1960.

There is, however, other evidence at the secondary school level that suggests a decline in standards during the 1960s and 1970s. This evidence is based on public examination performance. There has been considerable stability over the years in the percentage of students who pass such examinations. This might be interpreted as indicating a stability of standards. It would seem, however, more likely that the percentage uniformity is actually achieved at the expense of maintaining high attainment standards; that is, more students pass precisely because standards are lowered. While this may be so, the fact that the absolute number of students who achieve passes in public examinations (even if at a somewhat lower level of attainment) has increased steadily over the past 20 years is evidence of a general rise in the educational output of schools. Such a rise should lead, in time, to a general improvement in the educational standards of the nation as a whole.

REFERENCES

Bardell, G. S., Forrest, G. M., & Shoesmith, D. J. *Comparability in GCE. A review of the board's studies, 1964-1977.* Manchester: Joint Matriculation Board, 1978.

Brimer, A., Madaus, G. F., Chapman, B., Kellaghan, T., and Wood, R. *Sources of difference in school achievement.* Slough: NFER Publishing Company, 1978.

Burt, C. The mental differences between children. In C. B. Cox & A. E. Dyson (Eds.), *Black paper two. The crisis in education.* London: Critical Quarterly Society, no date.

Bush, P. Basic issues in awarding school leaving certificates. Comparability of grading standards in public examinations in England and Wales: Methods and Problems. In F. M. Ottobre (Ed.), *Criteria for awarding school leaving certificates.* New York: Pergamon, 1979.

Coolahan, J. Three eras of English reading in Irish national schools. In V. Greaney (Ed.), *Studies in reading.* Dublin: Educational Company, 1977.

Davis, F. B. Research in comprehension in reading. *Reading Research Quarterly,* 1968, *3,* 499-545.

Farr, R. *Reading: What can be measured?* Newark, Delaware: International Reading Association, 1969.

Forrest, G. M. *Standards in subjects at the ordinary level of the GCE, June 1970.* Manchester: Joint Matriculation Board, 1971.

Forrest, G. M. and Smith, G. A. *Standards in subjects at the ordinary level of the GCE, June 1971.* Manchester: Joint Matriculation Board, 1971.

Great Britain: Department of Education and Science. *Half our future.* A report of the Central Advisory Council for Education (England). London: Her Majesty's Stationery Office, 1963.

Great Britain: Department of Education and Science. *Progress in reading 1948-1964.* Education Pamphlet No. 50. London: Her Majesty's Stationery Office, 1966.

Great Britain: Department of Education and Science. *A language for life.* A report of the Committee of Inquiry appointed by the Secretary of State for Education and Science under the Chairmanship of Sir Alan Bullock F.B.A. London: Her Majesty's Stationery Office, 1973.

Great Britain: Department of Education and Science. *Primary education in England.* A survey by Her Majesty's Inspectors of schools. London: Her Majesty's Stationery Office, 1978.

Great Britain: Department of Education and Science Assessment of Performance Unit. *Mathematical development.* Primary Survey Report No. 1. [Report on the 1978 primary survey from the NFER to the Department of Education and Science and the Welsh office.] London: Her Majesty's Stationery Office, 1980.

Great Britain: Ministry of Education. *Reading ability.* London: Her Majesty's Stationery Office, 1950.

Great Britain: Ministry of Education. *Standards of reading 1948-1956*. Pamphlet No. 32. London: Her Majesty's Stationery Office, 1957.

Great Britain: Ministry of Education. *15 to 18*. A report of the Central Advisory Council for Education (England). Volume II: Surveys. London: Her Majesty's Stationery Office, 1960.

Great Britain: Departmental Committee. *Northern Ireland*. Final report of the Departmental Committee on the educational services in Northern Ireland, 1923-1924.

Harnischfeger, A. and Wiley, D. E. The marrow of achievement test score declines. In L. Lipsitz (Ed.) *The test score decline: Meaning and issues*. Englewood Cliffs, N. J.: Educational Technology Publications, 1977.

Horton, T. R. *The reading standards of children in Wales*. Slough: NFER Publishing Co., 1973.

Hunt, L. C. Can we measure specific factors associated with reading comprehension? *Journal of Educational Research*, 1957, *51*, 161-172.

Ireland. Dail Eireann. *Parliamentary Debates*, November 12, 1925. Vol. No. 13.

Ireland. Dail Eireann. *Parliamentary Debates*, May 28, 1941. Vol. No. 83.

Ireland: Royal Commission of Inquiry into Primary Education. Report of the commissioners (H. C. 1870, xxviii, pt. i).

Irish School Weekly, March 3, 1928.

Keenan, P. J. *Address on education to National Association for the Promotion of Social Science*. Dublin: The Queen's Printing Office, 1881.

Kelly, S. J., and McGee, P. Survey of reading comprehension. *New Research in Education*, 1967, *1*, 131-134.

Madaus, G. F., Airasian, P. W., & Kellaghan, T. *School effectiveness: A reassessment of the evidence*. New York: McGraw-Hill, 1980.

Madaus, G. F., Kellaghan, T., Rakow, E. A., and King, D. J. The sensitivity of measures of school effectiveness. *Harvard Educational Review*, 1979, *49*, 207-230.

Maxwell, J. *Reading progress from 8 to 15*. Slough: NFER Publishing Company, 1977.

McDonagh, D. G. A survey of reading comprehension in Dublin city schools. *Irish Journal of Education*, 1973, *7*, 5-10.

McGee, P. An examination of trends in reading achievement in Dublin over a ten-year period. In V. Greaney (Ed.), *Studies in reading*. Dublin: Educational Company, 1977.

McGee, P. Personal communication, 1980.

Peaker, G. F. Standards of reading of eleven-year-olds, 1948-1964. In Great Britain: Department of Education and Science, *Children and their primary schools. A report of the Central Advisory Council for Education (England)*. Volume 2: Research and surveys. London: Her Majesty's Stationery Office, 1967.

Start, K. B. and Wells, B. K. *The trend of reading standards*. Slough: NFER Publishing Company, 1972.

Travers, M. A second replication of a survey of reading comprehension in Dublin city schools. *Irish Journal of Education*, 1976, *10*, 18-22.

Ward, L. R. Measuring comprehension in reading. *College English*, 1956, *17*, 481-483.

Wilson, J. A. *Reading standards in Northern Ireland*. Belfast: Northern Ireland Council for Educational Research, 1973.

Willmott, A. S. *CSE and GCE grading standards: The 1973 comparability study*. London: Macmillan Education, 1977.

Willmott, A. S. & Nuttall, D. L. *The reliability of examinations at 16+*. London: Macmillan Education, 1975.

Chapter 9

Race, Social Class, Expectation, and Achievement

David T. Harris
Joseph L. Amprey, Jr.

The current general decline in Scholastic Aptitude Test (SAT) scores for all students can be attributed, at least in part, to a general decline in discipline in both the schools and the homes. Parents and educators both face a more rebellious group of young people and exert less authority over these youth than did the parents and educators of previous generations.

However, the biggest differences, with regard to academic achievement, have been and will probably continue to be caused by class differences. This is true in part because wealthier students experience smaller class sizes, better instruction, and many other advantages not experienced by poorer students. Black Americans are still subject to racism, and proportionately more blacks than whites are poor. Therefore, being black very often leads to lower academic achievement but not merely because of skin color.

Moreover, race still has a significant association with academic achievement—specifically with the SAT; those differences will be discussed in this chapter. This chapter will also include the recommendation that some nonacademic admissions criteria would make higher education more accessible and beneficial for black as well as all other Americans. We cannot and should not improve the quality or accessibility for one ethnic group without doing so for all. The barriers inhibiting blacks also inhibit poor underachieving whites as well as other minorities. By assisting some students we can assist all.

In recent times, racism and racists have been cited as the primary causes

The Rise and Fall of National Test Scores

Copyright © 1982 by Academic Press, Inc.
All rights of reproduction in any form reserved.
ISBN 0-12-068580-9

for minority student underachievement. However, it is believed here that educators, working directly with minority students, are capable of doing as much, if not more, damage than overt racists. Although racists may contribute to the generally lower expectations some people have of minorities, it is the covert low expectations of the educator who has direct contact with the student that are more influential.

Today, many black students are greatly influenced by these expectations. The greatest challenge facing us today is that of getting these "direct contact" educators to believe in the potential of black students.

Black students, particularly black males, are often considered unreachable and generally incompetent students. Lightfoot (1976) points out that "classroom research" describes the prototype image of black males as "hyperactive, disobedient, withdrawn and lazy," while white males are seen as "aggressive, and dominant, less likely to conform." White females are characterized as "perfect and obedient students," but Lightfoot notes that black females face both race and sex discrimination in the schools.

The perceived differences in the intellectual abilities of black and white students and the teacher's perception of the different behavorial characteristics provide a conceptual foundation for racial self-fulfilling prophecies. Some studies (Rubovits & Maehr, 1973) have found differences in the reinforcement of academic behavior between races in the classroom. Such studies would lend support to an argument that some educators do not expect black students to do well academically, and, thus, these expectations are fulfilled.

In studies concerned with "expectancy," there is another dimension that may influence the performance of black students; that is, What does the student himself expect his fate to be? Social learning theory tells us that we come to be what we are and that we develop a sense of self through a long and complex series of investigations with our environment. We develop a sense of how we fit into that environment and how we can and should interact in it. This is the essence of internal–external locus of control.

One is internally in control when one believes that a desired reinforcement can be elicited through one's own behavior. External control, however, is dominant in a person's belief system when it is expected that things will happen to him or her as a result of luck and that he or she has little control over fate (Rotter, 1966). The characteristics generally associated with internal control are positive. People who possess a strong internal locus of control think of themselves as powerful, active, achieving, independent and effective, while people who possess an external control belief think of themselves "in somewhat opposite terms" (Hersch & Scheibe, 1967). Many studies on internal–external control indicate that blacks, other minorities, and poor people are generally more externally controlled than are middle class people and whites (Rotter, 1966; Battle & Rotter, 1963; Lefcourt & Ludwig, 1965, 1966). Others have presented evidence that offers explanations for the appearance of an external locus of control in some black people. Gurin *et al.* (1969) explain that a phenomenon called "system blame" is occurring. Subjects, particularly in the Gurin study, were actually internally controlled—regarding themselves as the causal agent—when they responded externally to items that concerned people in society. The study indicates that this is a defense mechanism against racism in the environment of black people.

The evidence, almost regardless of how it is interpreted, leads us to one important conclusion. Most black students believe and expect that elements in their environment will respond differently toward them and other black people than those same elements will respond to others, and, in many cases, blacks will exercise little control over those elements. Furthermore, there is considerable evidence that these beliefs are well founded and that educators and others actually do respond differently to black students. The result is the establishment and maintenance of a system of differential expectation on the part of both educators and blacks themselves for black students. Educators will not expect black students to do as well as their white counterparts, and black students do not expect to be equitably reinforced for their work. Such, indeed, is the behavioral foundation of the "self-fulfilling prophecy." Thus, it is not too far-fetched to assume that performance on achievement and aptitude tests will reflect the cumulative effect of such experiences.

The major point to be made here is that the most devastating barriers occur over a 12-year (or more) period of time prior to college. Improved SAT scores and increased accessibility to and persistence in college for minorities must begin with changes that start at the nursery school level.

There are, in the meantime, some potentially useful changes that could be initiated by college personnel for improving minority access to and persistence in college. Experimentation with nonacademic variables in admissions and in guidance may be worthy of some further exploration. Several cluster personality characteristics which would seem worthwhile (e.g., motivation, social outlook, self-concept, and self-esteem) have been examined in professional literature. The social, mental, and physical adjustment of the student to the college or university, and vice versa, are also important. Likewise, the student's knowledge of the higher educational environment and of how to cope with it is extremely relevant. All of these factors may be as important as whatever can be demonstrated through standardized tests and high school records. However, these factors are usually not, in any significant way, considered as variables in admissions procedures at colleges and universities, and too little is known about the ways in which desirable personality traits can be enhanced for black students on an institutional basis during the precollege school experience.

Sedlacek and Brooks (1976) reviewed much of the literature about nonacademic correlates of success in higher education for black students. They concluded that there were seven variables that were not directly associated with academic achievement but could be very useful in predicting the academic performance of black students. These variables included a positive self-concept, understanding and dealing with racism, realistic self-appraisal, and preference for long-range goals over short-term or immediate needs. The others were the availability of a strong support person, successful leadership experiences, and demonstrated community services.

The black student who believed in himself and was confident that he or she could succeed was the more successful student. While most black students have the same problems and pressures of nonminority students, black students must also adjust to an environment which is often alien and hostile to them. The confident black student with a strong self-concept was found to be more successful in high school (Epps, 1969) and in college.

Closely associated with the positive self-concept is whether or not the student is able to understand and deal with racism. The black students who are able to analyze hostile and racist situations and develop behaviors to cope with such situations are inclined to be more successful. The Gurin study (1969) found that black students had conflicting systems of belief about being in control of their environments. While individuals believed that they could bring about a desired reinforcement through their own behavior, those same individuals also believed that people could not solely control their environment. There is, of course, a foundation for this construct in social reality, as the reward and reinforcement systems are probably more inconsistent for blacks than for whites due to racism. Those blacks who find ways to deal with such inconsistencies are more likely to make the necessary adjustments in college and succeed.

Another nonacademic correlate of success is whether the student has demonstrated service to his community. While this seems to be closely associated with successful leadership experience, black students who show an active interest in and involvement with their communities are also likely to develop innovative ways of coping with the particular circumstances which they are likely to encounter.

These seven constructs could be very useful in predicting the success of black students in higher education. It seems that these variables are also useful in providing information that might help identify support mechanisms for students after they have entered college. Each of these could contribute specifically to the retention rate of black students and, perhaps, to all students in general.

The extent to which these correlates are useful and what the practical application of these to the daily practices of admissions and academic guidance is has yet to be developed. A few institutions have tried to incorporate nonacademic correlates into their admissions policies, but it seems that the research and development of a standardized mechanism for using nonacademic correlates of success is yet to be designed. However, from the research, it seems evident that these variables could predict the performance of black students, and could suggest ways in which black students might be helped to deal more effectively with the lingering effects of racism. It is possible, even, that a significant portion of black students' test scores could begin to increase as a result of improved coping abilities.

Among the other correlates of success are the kinds of higher education institutions a student chooses to attend. For the millions of students who could benefit from selective institutions, there are hundreds, indeed thousands, of colleges with varying characteristics. Could a clear and accurate definition of the characteristics of colleges or universities, matched with a clear and accurate description of the types of students attending them, lead to a better, more appropriate, and perhaps more productive pairing of our student population with our colleges and universities? Clearly such improved choices would suggest better success in higher education (Arrow, 1974).

A variety of other correlates are also useful in predicting success of black students in higher education. The extent to which the student has learned and incorporated certain behaviors associated with academic success is yet another example. Is the student prompt? Does the student "appear" to understand academic subject matter? Does the student seek academic and other kinds of help when it is needed? Is the student aware of available resources, and does the student use them? These and more are all a part of important behaviors which can contribute to success in higher education.

The student who is able to analyze his own strengths and weaknesses is also more likely to succeed. This student can determine weaknesses in his preparation and take the necessary action for coping with them.

Students who can defer short-term goals for the greater gains of long-term goals are more likely to succeed in college. Black students who prefer the long-range goal of achieving a college education rather than trying to satisfy immediate needs for earning money, such as getting a job for minimum wage or going into the military service, are more likely to succeed in and finish college.

Students who have been exposed to strong people who can serve as models make better adjustments to higher education. Have schools experimented with making such models available for precollege black students, and have follow-up studies been done on college performance?

Successful leadership experience is another positive predictor of college success. Black students who have had such experience, particularly when it is demonstrated in nontypical circumstances, are likely to be successful in college. The black students who have been community or church leaders, or even street gang leaders, are cited as those who would be successful in college.

Although these nonacademic correlates have been primarily discussed here regarding the retention of black students, they should be just as effective when used with whites and other nonblack minorities.

Obviously, there is a need for more research in this area. Today, the retention of students is a major concern of most college and university administrators. If the exact influences of these nonacademic correlates could be operationally delineated, the admissions and retention processes would be significantly enhanced. Perhaps, as a result of definitive research, certain personality assessment instruments might be included in the admissions process.

SUMMARY

Although race has been identified as a major component in the debate on declining test scores, the general decline in discipline and natural class differences influence minority test scores more significantly than does race per se.

The educators who have direct contact with students are the persons who possess a very large share of the influence on student achievement. These "direct contact persons" are stifled mostly by low, negative expectations of their pupils. One of the greatest challenges facing education today involves ways to change the expectations of educators who have direct contact with students. These changes need to begin at the nursery school level.

Finally, a worthy challenge for admissions officers and guidance personnel might be the initiation of experimentation with certain nonacademic variables that may play a large role in predicting college student achievement. These variables should be applicable for all students, regardless of socioeconomic level or ethnicity.

REFERENCES

Arrow, K.J. *The Limits of Organization,* New York: Norton, 1974.
Battle, E. and Rotter, J.B. Children's feelings of personal control as related to social class and ethnic group. *Journal of Personality,* 1963, *31,* 482–490.
Epps, E.G. Correlates of academic achievement among northern and southern urban negro students. *Journal of Social Issues,* 1969, *25,* 5–13.

Gurin, P., Gurin, G., Lao, R.C., and Beatie, M. Internal–external control in the motivational dynamics of negro youth. *Journal of Social Issues,* 1969, *25,* 29–53.

Hersch, P., and Scheibe, K. Reliability and validity of internal–external control as a personality dimension. *Journal of Consulting Psychology,* 1967, *31*(6), 609–613.

Lefcourt, H.M., and Ludwig, G.W. The American negro: A problem in expectancies. *Journal of Personality and Social Psychology,* 1965, *1,* 377–386.

Lefcourt, H.M., and Ludwig, G.W. Alienation in negro and white reformatory inmates. *Journal of Social Psychology,* 1966, *68,* 153–157.

Lightfoot, S.L. Socialization and education of young black girls in schools. *Teachers College Record,* 1976, *78,* 239–262.

Rotter, J.B. Generalized expectations for internal versus external control of reinforcement. *Psychological Monographs: General and Applied,* 1966, 80, No. 1 (whole no 609).

Rubovits, P.C., and Maehr, M.L. Pygmalion in black and white. *Journal of Personality and Social Psychology,* 1973, *25,* 210–218.

Sedlacek, W.E., and Brooks, G.C., Jr. *Racism in American education: A model for change.* Chicago: Nelson-Hall, 1976.

Chapter 10

The Final Hurdle:
Minimum Competency
Achievement Testing

Richard M. Jaeger

WHAT IS MINIMUM COMPETENCY ACHIEVEMENT TESTING? AND WHY?

"The State Board of Education shall adopt tests or other measurement devices which may be used to assure that graduates of the public high schools and graduates of nonpublic high schools supervised by the State Board of Education . . . possess those skills and that knowledge necessary to function independently and successfully in assuming the responsibilities of citizenship. This Article has three purposes: (i) to assure that all high school graduates possess those minimum skills and that knowledge thought necessary to function as a member of society, (ii) to provide a means of identifying strengths and weaknesses in the education process, and (iii) to establish additional means for making the education system accountable to the public for results [General Assembly of the State of North Carolina, Article 39A, Section 115-320.6 'Purpose,' 1977]."

In a nutshell, that's what minimum competency achievement testing is in North Carolina. In keeping with further provisions of the law, the State Department of Public Instruction administers a reading test and a mathematics test to all eleventh graders in the state. If after four attempts to pass these tests a student fails to do so, the student receives a "certificate of completion" in lieu of a high school diploma.

The definition of minimum competency testing, in intent and operation,

223

The Rise and Fall of National Test Scores

Copyright © 1982 by Academic Press, Inc.
All rights of reproduction in any form reserved.
ISBN 0-12-068580-9

is somewhat different in some other states; however, North Carolina's definition is not unusual. The National Institute of Education recently contracted with National Evaluation Systems, Inc. (NES) to conduct a review of minimum competency testing programs throughout the nation. After consulting directors and staffs of competency testing programs in 31 states and in 20 school systems, and following an extensive review of the ever-burgeoning literature on minimum competency testing, NES concluded that "there is no consistent terminology for minimum competency testing in use in the testing field [Gorth & Perkins, 1979a, p. 8]." The NES document cites nine definitions of minimum competency testing found in the research and education policy literature. Five of these attempt to specify the essential elements of a minimum competency testing program, and the others speculate on the likely consequences of such a program. Among the more specific definitions we find:

> [Minimum competency testing programs are] "organized efforts to make sure public school students are able to demonstrate their mastery of certain *minimum* skills needed to perform tasks they will routinely confront in adult life [American Friends Service Committee, 1978]."
> "Minimum competency tests are constructed to measure the acquisition of competency or skills to or beyond a certain defined standard [Miller, 1978]."
> [Minimum competency testing programs are] "testing programs which attempt to learn whether each student is at least 'minimally competent' by the time the student graduates from public school [National School Boards Association, 1978]."
> [Minimum competency testing is] "a certification mechanism whereby a pupil must demonstrate that he/she has mastered certain minimal skills in order to receive a high school diploma [Airasian *et al.,* 1978]."
> "Nearly all minimum competency testing programs seek to define minimum learning outcomes for students in a variety of academic areas and to insure that these standards are satisfied [Cohen & Haney, 1980]."
> "Minimum competency testing involves the administration of proficiency tests in order to certify that minimum competency or proficiency exists with regard to a well-defined set of knowledge or skills [Beard, 1979, p. 9]."

If the reader is suffering from more than a bit of circularity and from an absence of clarity at this point, the intended outcome has been achieved. Although proponents of minimum competency testing appear to agree on the intended benefits of the tests, there is decided lack of agreement on the defining characteristics of such programs. Two features, however, appear to be common to all minimum competency testing programs:

1. Test results are used to make decisions about the educational future of individual students.
2. The programs incorporate some methods of deciding which students have been successful and which students have failed.

These two characteristics of minimum competency testing distinguish such programs from many other testing activities that pervade elementary and secondary education in the United States. Achievement testing in reading, mathematics, and other basic skills subjects is certainly not new. Five years prior to the Florida legislature's 1971 act mandating the development of the first statewide minimum competency testing program, it was estimated that over 150 million test booklets were sold to U.S. schools. In that year (1966), the enrollment in our elementary and secondary schools was about 50 million. Hundreds of thousands of public school teachers dutifully peeled millions of gummed labels from score reports returned by testing companies and pasted them in millions of students' cumulative record folders to be filed and forgotten unless an atypical parent asked about a child's scores. With minimum competency testing, however, the "file and forget" strategy is not viable. The programs demand that important decisions—promotion to a new grade, the awarding of a high school diploma, assignment to a remedial program, etc.—be based on the test scores. The programs demand that pass–fail decisions be made, and these decisions in turn require that specific standards be established. Unlike typical achievement testing programs, minimum competency testing programs do not allow performance to be interpreted solely in terms of relative achievement, nor do they provide data on only the institutional performance of schools, of school systems, or of state-wide instruction—as is the function of many statewide assessment programs.

Why the sudden development of minimum competency testing programs? The most obvious explanation is that minimum competency testing is a logical outgrowth of the increasing public clamor for accountability in the schools and of the public's acceptance of a factory model of schooling (Callahan, 1962). In an attempt to place competency testing in the larger context of U.S. social policies, Cohen and Haney (1980) point to our longstanding interest in having government promote minimum levels of social welfare. Public health programs, social security, unemployment insurance, welfare programs, and free public education are cited as examples of such policies.

In contrast to these earlier programs, minimum competency testing demands the maintenance of minimum levels of outcomes, rather than minimum levels of inputs—such as minimum teaching time per pupil. This is consistent with the growing public realization that social welfare programs don't necessarily buy effective solutions. Social security is demonstrably insufficient to protect older citizens from the ravages of inflation. Welfare programs appear to beget increasing numbers of clients, and the corresponding work of social service agencies appears to be largely ineffective in restoring welfare recipients to self sufficiency. The strong relationship between socioeconomic disadvantage and educational deficit, widely publicized in the works of Coleman *et al.* (1966) and Jencks *et al.* (1972), persists despite massive infusions of federal funds through such programs as Head Start, Title I of the Elementary and Secondary Education Act of 1965, the Emergency School Assistance Act, and dozens of lesser programs.

Statistics on student achievement in secondary schools, well-documented in other chapters of this book, have not been lost on the public. In "College Entrance Examination Trends" (Chapter 2, this volume), Eckland documents the decline of Scholastic Aptitude Test scores since the early 1960s, and in Chapter 6 "Trends in Academic Performance in Arithmetic from Primary Grades through High School," Fey and Sonnabend illustrate the decline of test scores in mathematics since 1963. In far less detail, and without the careful explanatory analyses presented in this volume, basic trends in achievement test scores have been well reported in the popular press. Coupled with frequent reports on the increasing costs of education, these data have led to a loss of faith in education. That is, the public is no longer satisfied to rely on the schools, and those who run the schools, to provide students with minimally sufficient academic skills.

This view of the genesis of minimum competency testing is supported by the opinions of a number of commentators in addition to data compiled in the NES review of minimum competency testing programs currently in operation (Gorth & Perkins, 1979a). Beard (1979) states:

> [Minimum competency testing] has widespread popular appeal to citizens and politicians who see it as a way of holding schools accountable and forcing them 'back to the basics.' These groups are convinced that the quality of public education has been eroded over a period of years and that high schools are graduating significant numbers of students who are unable to read and write, and, consequently, unable to support themselves through gainful employment [p. 9].

The views of Lott (1977) are thematically similar:

> There are probably two reasons for the current interest in basic competency testing throughout the nation. First, there is concern that schools are awarding high school diplomas to individuals who lack some of the basic skills, who can't read and write, for example. In addition, because of the cost of education, there is continuing interest in accountability [p. 2].

The summarization of data provided in NES's final project report (Gorth & Perkins, 1979b) shows that legislative action, rather than initiative on the part of educators, was responsible for the development of minimum competency testing programs in 14 of the 31 states investigated. Action by a state board of education was cited as the locus of an initial competency testing mandate in 16 other states. In interpreting these data, it should be noted that state boards of education are typically composed of lay members rather than professional educators and that state boards often act so as to forestall the impending actions of state legislatures.

Further support of the contention that minimum competency testing programs constitute political responses to perceived public demand is provided by the existence of, and widespread media attention to, educational "malpractice" suits on behalf of plaintiffs who, despite having spent the usual 12 years in the public schools, claimed not to have received an adequate education (*Peter W.* vs. *San Francisco Unified School District,* 1976; *Donahue* vs. *Copaigue,* 1977).

In sum, these observations suggest that minimum competency achievement testing is primarily a political movement, not an educational movement. Such programs reflect the blind faith of state legislatures and state boards of education in their power to mandate—through law, regulation, or administrative action—some minimum level of educational success. The faith of the lawmakers and board members in the efficacy of minimum competency testing has often been exemplified in a series of assertions compiled by Gorth and Perkins (1979a). They found that proponents of minimum competency testing believe it will "(1) restore confidence in the high school diploma, (2) involve the public in education, (3) improve teaching and learning, (4) serve a diagnostic, remedial function, and (5) provide a mechanism of accountability [p. 12]."

The likelihood that these valued ends can be achieved through minimum competency testing programs is examined below.

THE GROWTH OF MINIMUM COMPETENCY TESTING PROGRAMS AND THE USE OF TEST RESULTS

Although a number of researchers agree that minimum competency achievement testing programs exist in a majority of the states, they disagree on the exact number currently in operation. Baratz (1980) and Gorth and Perkins (1979b) list 31 states as having some form of minimum competency testing program. However, Alaska, Colorado, Indiana, Oklahoma, Washington, and Wyoming are contained in Baratz's list but do not appear in the Gorth and Perkins list, whereas Illinois, New Hampshire, and South Carolina are contained in the Gorth and Perkins list but are not contained in Baratz's list. Combination of the two lists thus leads to the conclusion that some form of minimum competency testing program exists in at least 35 states as of this writing.

Both Baratz (1980) and Gorth and Perkins (1979b) list the year of enactment of legislation or state board action that mandated minimum competency testing programs in most states. From their combined data, it appears that only two states had mandated minimum competency testing programs as early as 1971, that four more took similar action in 1972, but that none were added during 1973 and 1974. In 1975, five more states enacted minimum competency testing programs, and four more were added in 1976. Thus, by the end of 1976, some 15 states had some form of minimum competency testing requirement. The rapid acceleration of the movement can be noted from data for 1977, when an additional nine states mandated competency testing programs, and for 1978, when the ranks of states with competency testing programs were increased by eight. In this two-year span, the number of states with minimum competency testing programs more than doubled. Unfortunately, data on actions during 1979 could not be found.

The required or intended uses of minimum competency testing data are often specified in enabling legislation or in the actions of state boards of education. Gorth and Perkins (1979b) list five uses that occur most frequently: (a) determining which students will receive high school diplomas; (b) providing special recognition of achievement on high school diplomas; (c) making decisions on students' promotions to the next grade; (d) identifying students for mandatory remediation; and (e) identifying students for optional remediation programs. Of these uses, determining which students will receive high school diplomas is specified most frequently. Combining data provided by Baratz (1980) and by Gorth and Perkins (1979b), 20 states can be identified as using competency test scores for this purpose. Twenty-five states use competency test scores either to assign students to mandatory remedial instruction programs or to identify students for optional remedial instruction. Competency test scores are considered in grade promotion decisions only in three states, and only three states use competency test scores as a basis for indicating special recognition of achievement on high school diplomas.

CHARACTERISTICS OF COMPETENCY TESTING PROGRAMS, WITH TWO EXAMPLES

Although discussions of minimum competency testing programs often suggest that the label represents a unitary, well-defined set of activities, such is not the case. Minimum competency testing programs differ in important ways despite their universal adherence to standards of performance and their impact on individual students. Among other dimensions, it is possible to describe competency testing programs in terms of the agency that controls them (state departments of education or local school systems), the types of competencies or skills emphasized in the tests (basic academic skills or so-called "life skills"), the breadth of subject matter included in the tests, the grade levels at which testing takes place, and the types of standards the tests impose (standards defined in terms of a total test score, standards defined in terms of several subtest scores, or standards defined in terms of performance on competencies, objectives, or sets of test items). Gorth and Perkins (1979b) summarize information on these and other testing program characteristics for the state and school district competency testing programs their study reviewed. We will turn, shortly, to examples that illustrate how these dimensions come into play in actual practice.

One of the most important determiners of the political impact of a competency testing program is the locus of control of the program. In 24 of the states Gorth and Perkins investigated, the state department of education determined the primary emphases and content of competency tests and imposed unitary, statewide standards as well. In 7 states, the state department of education had no part in selecting competency tests nor in setting standards. The state merely required that local school systems assume responsibility for both of these functions. It has long been claimed that external testing programs have a profound impact on the content, structure, and sequence of the school curriculum and on the methods teachers use for instruction. Therefore, the locus of control of a minimum competency testing program has far-reaching implications for the schools.

As examples of the important differences among minimum competency testing programs, we shall consider two such programs in detail. The program in North Carolina is typical of those operating in the majority of competency testing states. In North Carolina, the state department of education maintains a high degree of control. The second program to be considered is the one in Oregon, where local school systems determine the competencies to be examined, the methods whereby students demonstrate their competence, and the standards to be used for demonstration of competence.

North Carolina's Minimum Competency
Achievement Testing Program

The fundamental purposes of the minimum competency testing program operating in North Carolina are defined in the enabling legislation cited earlier. First, the legislature believed that high school graduates must possess some definable body of knowledge and skill in order to "function independently and successfully" in society. The existence of a definable set of survival skills was thus assumed in the legislation. The legislature also assumed that a minimum competency testing program would ensure that all high school graduates awarded a diploma in North Carolina possessed these necessary survival skills. Second, the legislature specified that the minimum competency testing program serve a diagnostic purpose. The tests or other measures adopted for use in the program were to enable the state's educators to "identify strengths and weaknesses in the education process." Finally, the program was specifically intended to impose additional state-level control on public education in North Carolina by "making the education system accountable to the public for results."

To accomplish these various purposes, the minimum competency testing law established a High School Competency Test Commission composed of public school teachers and administrators, members of the lay public, two specialists in psychometrics, and two university faculty members. The commission was to recommend to the State Board of Education a set of measurement instruments to be used in assessing students' competencies and was to advise the board on required standards of performance. As the program has operated thus far, the State Board of Education has uniformly adopted the recommendations of the competency test commission.

The first administration of competency tests was to be for research purposes only. This enabled the commission to gain information that would help it formulate recommendations on selection of measures and establish competency standards. This provision of the law was, perhaps, the program's most enlightened feature. The first administration of tests was to take place in the spring of 1978, and, following an evaluation of the results, the commission was to make its recommendations to the State Board of Education on tests and standards to be used in operational competency testing of all eleventh graders, beginning in the fall of 1978.

After a series of hearings and consultations with persons who had been responsible for the development of competency testing programs in other states, the commission decided that the kinds of skills necessary to survival in society involved the "functional application" of reading and mathematics. They searched for measures that would assess students' abilities to "apply these skills to practical situations [Gallagher, 1980]."

To define the competencies and skills involved in the functional application of reading and mathematics in greater detail, the commission developed a list of more than 250 objectives for minimum competency in those areas. Among the reading objectives were "to draw inferences from various materials," and "to select the main idea and related details from various passages." In mathematics, objectives included "to compute using whole numbers," and "to interpret and use maps, graphs, charts, and tables." Lists of objectives were sent to all local school systems in North

Carolina, and these schools were asked to have their reading and mathematics teachers place the lists in priority order.

The commission reviewed some 15 commercially available competency tests in reading and mathematics. With the advice of teachers and other consultants, the commission identified the three tests in each area that, in its judgment, assessed the high priority competency objectives identified by the state's teachers. The three reading tests selected by the commission were administered to half of the eleventh graders in the state, together with a norm-referenced achievement test in reading. The three mathematics tests selected by the commission were administered to the other half of the state's eleventh graders, also together with a norm-referenced achievement test.

Results from the competency tests administered for research purposes were used to select and modify competency tests for operational use and to establish recommended standards of performance. The commission obtained the advice of a small sample of teachers on both of these issues. It also reviewed the results of a variety of analyses of the performance of eleventh graders on individual test items and on entire tests. The details of these analyses are discussed by Gallagher (1980).

As the North Carolina High School Competency Testing Program currently operates, all eleventh graders in the public schools take a reading test and a mathematics test in the fall term. If they answer at least 72% of the items on the reading test correctly and at least 64% of the items on the mathematics test correctly, they are eligible for high school diplomas. To be awarded diplomas, students must also complete all of the attendance and class requirements typically required for high school graduation. If they fail either or both of the competency tests, they are given three additional opportunities to pass them prior to the spring term of their twelfth-grade year in school. If a student has not passed both the reading test and the mathematics test after four attempts, he or she is given a "certificate of completion" in lieu of a high school diploma. Students who do not receive regular high school diplomas because they have not passed the competency tests can continue to take the tests each year following the graduation of their high school class, until they pass the tests or until they reach the age of 21. Upon passing the tests, students can exchange their certificates of completion for regular high school diplomas.

The competency testing law requires that the schools offer remedial instruction to those who have failed the competency tests, including those beyond the normal age for high school graduation.

Although this brief review of the development and operation of competency testing in North Carolina suggests that the High School Competency Test Commission engaged in extensive consultation with local school system personnel in reaching decisions on which competencies to assess, on tests to use, and on standards to employ, it is clearly the case that the ultimate decisions on these issues were made by the commission and by the State Board of Education. Thus, in North Carolina, it is the state that imposes external criteria of satisfactory performance on all high school students, and, consequently, on all local school systems.

Oregon's Minimum Competency Achievement Testing Program

In December of 1974, the Oregon State Board of Education adopted a guide containing revised minimum standards for the public schools of the state. One component of the minimum standards defines new high school graduation requirements for all public schools in the state. Oregon's version of high school competency testing was developed in response to the new graduation requirements.

Herron (1980) presents a detailed history of the development of the revised standards for schools and of Oregon's attempt to establish competency-based graduation requirements that were consistent with the standards. The path he describes is politically treacherous and fraught with conflicts between idealistic rhetoric and the real limits of the measurement and assessment art. This brief review borrows heavily from his report, and is restricted to a few salient highlights.

In contrast to the program of competency testing adopted in North Carolina, Oregon's State Board of Education adopted broad goals for education in the state and then required that local school systems develop more specific graduation competency requirements, develop and adopt procedures through which students could demonstrate their competence, and determine appropriate competency standards.

The administrative rules promulgated by Oregon's State Board of Education have the force of law, since the Oregon legislature has delegated statutory authority to the board to issue such rules. In effect, then, the State Board of Education established legal requirements that local school boards "shall award a diploma upon fulfillment of all state and local district credit, competency and attendance requirements," and further, that "the local board may grant a certificate identifying acquired minimum competencies to students having met some but not all requirements for the diploma and having chosen to end their formal school experiences [Oregon Administrative Regulation 581-22-228]."

The language of the Oregon regulations is quite specialized, including such terms as "program goals," "district goals," "performance indicator," "planned course statement," and so on. In total, 38 specialized terms are defined in the regulations concerning graduation requirements. Using such language, the State Board of Education required all local school districts to "establish minimum competencies and performance indicators beginning with the graduating class of 1978; certify attainment of competencies necessary to read, write, speak, listen, analyze and compute beginning with the graduating class of 1978; certify attainment of all competencies beginning not later than with the graduating class of 1981. [Oregon Administrative Regulation 581-22-236]." A performance indicator is defined to be "an established measure to judge student competency achievement," and a competency is "a statement of desired student performance representing demonstrable ability to apply knowledge, understanding, and/or skills assumed to contribute to success in life role functions [Oregon's Administrative Regulation 581-22-200]."

By 1981, all districts are required to record on high school students' transcripts whether or not the students have demonstrated competencies necessary to:

1. use basic scientific and technological processes
2. develop and maintain a healthy mind and body
3. be an informed citizen in the community, state, and nation
4. be an informed citizen in interaction with environment
5. be an informed citizen on streets and highways
6. be an informed consumer of goods and services
7. function within an occupation or continue education leading to a career [Oregon Administrative Regulation 581-22-231]

The board left the local district in charge of identifying specific competencies associated with each of these grandiose goals, developing performance indicators for each competency, and establishing standards of competence. The board did support six pilot projects in local districts that produced samples of program goals, competency statements, and performance indicators in three goal areas. As Herron (1980) points out, the materials produced by these projects were not consistent with the "life skills" orientation of the board's statewide graduation requirements. Despite this inconsistency, the pilot project materials were widely disseminated and became models for competency demonstration in many local school districts. The pilot project materials emphasized competencies defined in terms of traditional subject-matter skills, and performance indicators often involved traditional paper-and-pencil tests. For a number of competencies, classroom teachers were advised to observe students' traditional coursework and certify students' competence on the basis of those observations.

The practical problems resulting from Oregon's competency-based graduation requirements are well documented in Herron's (1980) report.

Effective measures of students' career development skills and social responsibility are generally unavailable. Some teachers are unwilling to run the risk of personal liability in certifying that students possess vaguely defined competence within elusive domains. Although some local districts have attempted to embrace the "life skills" goals embodied in statewide graduation requirements, others have adhered firmly to more familiar, basic school skills. The resulting inconsistencies from school district to school district have created massive problems for transfer students and raised the spectre of unequal educational opportunity throughout the state. Herron concludes, nonetheless, that Oregon is unlikely to adopt a statewide minimum competency testing program of the sort now underway in North Carolina and many other states.

In summary, the competency testing programs operating in North Carolina and Oregon exemplify the wide range of such programs. North Carolina's program is clearly controlled by the state. It involves uniform specification of competencies, measures, and standards. The state develops and distributes all tests, the state scores students' performances, and the state provides local school districts with information on which students have passed and which have failed. The North Carolina program is very limited in subject-matter scope.

In contrast, the program in Oregon is, within very broad state guidelines, controlled by local school districts. Districts specify competencies, identify indicators of competent performance, and establish standards. All measurement and reporting functions are handled by local school districts. The subject-matter scope of the program is extremely broad, and goes well beyond the traditional curriculum of the schools. Whether either program has positive effects on secondary education cannot be determined at this early stage in their operation. There are, however, obvious grounds for concern.

SOME LIKELY CONSEQUENCES
OF MINIMUM COMPETENCY TESTING

Whether, as their advocates suggest, minimum competency testing programs will reverse the current downward trend in achievement test scores, make the schools accountable to the public, restore meaning to the high school diploma, and rid the nation of functionally illiterate high school graduates, remains to be seen. It is already clear, however, that the achievement of these ends through minimum competency testing, were they achieveable, would not be without costs. Minimum competency achievement testing is a vast social experiment. It will surely affect the organization and governance of the schools, the school curriculum and its associated instructional procedures, the roles and functions of classroom teachers, the opportunities and experiences of students—both in school and in later life, and relationships between the schools and other social institutions. Logical analysis, experience with earlier testing programs, and preliminary data on the outcomes of competency testing provide some basis for predicting its ultimate consequences. Some of those predictions will now be reviewed.

Minimum Competency Testing and the High School Curriculum

When we speculate on the effects of minimum competency testing on the curriculum of U.S. high schools, three questions appear relevant. First, Is there any evidence that the use of external (to a school system) testing has any effect on the content or structure of the schools' curriculum? Second, Is there sufficient content similarity in the currently used minimum competency achievement tests to suppose that they could influence the curriculum of U.S. high schools in some uniform way? And third, Is the content of currently used minimum competency achievement tests materially different from the content of the curricula offered by U.S. high schools, or from alternative curricula that are judged to be appropriate or necessary on the basis of some external criteria? It would seem that all three of these questions must be answered in the affirmative if we are to conclude that minimum competency achievement testing is likely to have substantial curricular impact and that this impact is likely to be worrisome.

Although it has been claimed that minimum competency achievement testing is a relatively new phenomenon, it must be realized that tests have long been used to establish external criteria for high school graduation in some locales. The Boston public schools made use of a common examination for certification of high school students' proficiencies as early as 1845, and the New York State Regents Testing Program has antecedents dating back to the late 1870s. The stated purpose of these examinations was "to furnish a suitable standard of secondary school graduation." In documenting the effects of these early examinations on the curricula of New York high schools, Perrone (1979) states:

> The evidence is that the tests influenced significantly what was taught. The diaries of early twentieth-century teachers were filled with accounts of the long periods in which they prepared students for the state examinations, giving up in the process what they considered to be more engaging for the students. The *Regents Inquiry Into the Character and Cost of Public Education in New York State* reported in 1936 that the Regents Examination had, in effect, become the curriculum [p. 5].

Perrone goes on to suggest that both teachers and students used the tests to define what was worth teaching and what was worth learning. Although local school systems had curriculum goals that were far broader than the topics tested by the regents examinations, these goals were virtually ignored in the day-to-day activities of the schools.

The power of tests to determine what is taught has not gone unnoticed by measurement practitioners and psychometricians. In 1951, E.F. Lindquist suggested that "the contribution of educational measurement to education generally depends as much or more upon *what* test constructors elect to measure as upon how well they measure whatever they do measure [p. 140]."

Further evidence on the influence of test content on curriculum content is available in the international studies reviewed by Madaus and Airasian (1978). They found that where external certifying examinations were used, students and teachers tended to focus class instruction on objectives that could be gleaned from the content of certifying examinations used in the past, rather than on objectives listed in course outlines. The finding was quite general, spanning studies conducted in different nations and in different decades. Madaus and Airasian concluded "most studies have found that the proportion of instructional time spent on various objectives was seldom higher than the predicted likelihood of their occurrence on the external examination."

The minimum competency testing program in Florida and, particularly, the Functional Literacy Test used by that state, have been subject to recent review by a team of researchers including curriculum theorist Ralph Tyler. The program also has the distinction of being the object of litigation in federal court. In commenting on the apparent influence of Florida's Functional Literacy Test on the high school curriculum, Tyler (1979) stated, "We were told that many teachers interpreted the emphasis on basic skills to mean that they must devote most of their attention to routine drill [p. 30]." In testimony heard during the federal court case, it became apparent that the content of Florida's minimum competency tests largely determined the detailed curriculum of the remediation programs that are a part of Florida's minimum competency testing program. For students who failed the test repeatedly, and who were therefore assigned to remediation programs for a large portion of their school day, test material defined the curriculum.

Some might grant that externally imposed tests have substantial influence on the curriculum of the schools, but argue that such influence is both appropriate and valuable. To sustain such an argument, it would have to be agreed that the content of the tests constituted an appropriate curriculum for the schools. However, review of the minimum competency achievement tests that are currently in use suggests otherwise.

In their analysis of minimum competency testing programs in thirty-one states, Gorth and Perkins (1979b) found that the competency tests of 14 states emphasized academic skills exclusively, while the tests used in 16 other states emphasized a combination of academic skills and life skills. Such apparent breadth might be praised, since many have argued for greater "relevance" in the secondary school curriculum and for an easing of the transition from school to work. However, in more than half of the states, the specific content coverage of competency tests used at the secondary school level includes only reading, mathematics, and writing. Moreover, many of these tests demand nothing more than recognition of basic subject-matter mechanics or the application of basic mechanics to so-called life-skills situations. For example, an item on the test that one state uses in mathematics lists prices of various foods and drinks on a menu, and requires examinees to *recognize* the correct sum of prices for three menu items from among four listed sums. Writing tests used in minimum competency testing programs are often in multiple-choice form, and require nothing more than recognition of appropriate grammatical forms. Although such skills might be judged important by many educators, it would be difficult to justify substituting them for the traditional high school curriculum.

Broudy (1980), citing his earlier work with Smith and Burnett (1964), stated that a general education curriculum should include at least five strands of content: (*a*) skills in reading, writing, computation, and interpretation of aesthetic clues; (*b*) the basic sciences of mathematics, biology, chemistry and physics; (*c*) developmental studies in the earth sciences and in the social institutions and culture of human beings; (*d*) individual and collective problem solving; and (*e*) studies in the humanistic classics of the culture as exemplars of knowledge and value. Broudy suggests that these strands of content, when taken together, can aid students in building conceptual systems and networks of images that will "supply educated contexts for problem solving, feeling, judging, and communicating." He asks, rhetorically, "Do all children 'need' this kind of curriculum?" "No," he answers, "only children who are to become 'educated' adults do [p. 111]."

Reflecting on the sparse interpretation of functional literacy evidenced by the content of many minimum competency tests, Broudy (1980) asserts:

> Mechanical identification of printed words with their phonetic equivalents and their standard referents is not what is ordinarily meant by reading comprehension, let alone *functional* literacy. Whoever doubts this conclusion need only hand a non-English-speaking reader a dictionary and ask him to be functionally literate about the locution: "They worked around the clock." The point is that *other strands of the curriculum* are needed to provide context-building resources that make literacy possible, in save the barest mechanical sense. If acquisition of these resources is restricted, even intensive instruction in the mechanics will not produce literacy [p. 113].[1]

---

[1] Copyright 1980, American Educational Research Association, Washington, D.C.

In summary then, it is apparent that externally imposed tests have substantial influence on the curriculum of the schools. For those who have the greatest difficulty passing the tests, the content of the tests may *become* the curriculum. As most state-adopted minimum competency tests are presently constituted, the mechanics of the three R's or the simple application of mechanics to objects that may be encountered in adult life (grocery bills, checking accounts, newspaper advertisements, etc.) define test content. It is widely accepted by curriculum theorists, if not by constructors of minimum competency tests, that a curriculum consisting solely of language mechanics and the mechanics of computation cannot produce functionally literate high school graduates, for the term itself suggests the ability to succeed with the language demands of the adult world (Broudy, 1980; Amarel, 1980).

The Effects of Competency Testing
on Teachers and Students

Traditionally, responsibility for success in schooling has been shared by students, teachers, and school administrators. Achievement has been measured in terms of credits earned, courses completed, and years of attendance. Although it has been the responsibility of students to participate in classes and other school activities, both teachers and administrators have shared responsibilities for defining the nature and content of those activities and for providing instructional offerings that were consistent with their perceptions of students' needs. In several ways, minimum competency testing conflicts with these traditions.

Basing his analyses on the theoretical framework proposed by Ryan (1971), Blau (1980) suggests that minimum competency testing is a classic case of "blaming the victim." By creating a singular penalty for ultimate failure to pass a minimum competency test, the student is denied a high school diploma. The educational system, including teachers, curriculum supervisors, counselors, and school administrators, is excused from its responsibility for the student's failure. Minimum competency achievement tests are used to examine the performance of students as individuals, not to examine the performance of schools as institutions. Both the rewards for success and the penalties for failure associated with competency testing are imposed on individuals. This policy suggests that students alone are to be held accountable for their competency test performance regardless of the opportunities they have or have not been afforded to learn what is tested. In this regard, Blau states:

> Under the guise of "helping" the victim, it is necessary to examine the victim carefully, scientifically, objectively, mathematically, and so forth. The purpose is to confine the solution of the dilemma to manipulations of the victim. By concentrating all effort on the victim, it is possible to

displace, ignore, and ultimately avoid the basic social causes of the problems being addressed. The most important subtle effect of this ideological process is that when one concentrates on the victim, one can avoid passing judgment on one's own adequacies. To concentrate on the inadequacies of our students in the school system is to sanctimoniously imply a "not guilty" verdict for ourselves [p. 175].[2]

All this is not to suggest that students should be held blameless for their failures, nor that academic rewards should be based solely on effort rather than achievement. It is to suggest that the tradition of shared responsibility for academic success and failure is a valid and appropriate one. There are many causes of student failure, only some of which are within the power of students to control. Until such time as school systems and state governments can demonstrate that they should be held blameless, it seems inherently unjust to penalize students alone for their failure to succeed on externally imposed tests.

Although at least one teachers' union has taken a strong stand against minimum competency achievement testing (McKenna, 1979), there has been no evidence of a major anti-competency-testing movement on the part of classroom teachers. Indeed, an informal survey of high school teachers in one state (Jaeger *et al.*, 1980) revealed that 28% "strongly agree" and an additional 32% "agree" with the statement "I favor the minimum competency testing program as it is currently operating in this state." The state used for the survey was North Carolina and, as discussed above, North Carolina's competency testing program is strictly a state operation.

Despite teachers' apparent complacency on competency-testing issues, a number of researchers and commentators outside the classroom have suggested that teachers *should* be very concerned. Others have suggested reasons why teachers might accept, or even welcome, competency testing.

In the minds of many, failures of American education are viewed as failures of classroom teachers (Reilly, 1978; Hentoff, 1978). Bardon and Robinette (1980), citing the judgments of Hart (1978) and Fremer (1978), suggest that competency testing might provide a convenient escape hatch for teachers, and might actually reduce their vulnerability to public pressure. If the public views competency testing as a device for getting the schools back to basics, as a means of imposing rigorous achievement standards, and as a guarantee that schools will no longer graduate functionally illiterate students, the mere existence of competency testing programs might satisfy the public desire for educational accountability—at least for a time. And, since failing a competency test is most often treated as a student's problem, and not the school's problem, teachers are further relieved of responsibility. It is not surprising that teachers might welcome a lowering, or at least a redirection, of the heat.

---

[2] Copyright 1980, American Educational Research Association, Washington, D.C.

In addition, the existence of competency testing might communicate to some otherwise cynical students that the schools now mean business. No performance means no diploma. Competency tests might then become another weapon in the teacher's arsenal of devices for keeping students in line and on the task of learning (Gentry, 1976). Whether these instructional and political benefits of competency testing actually accrue to teachers is yet to be seen. Nonetheless, it is conceivable that teachers might expect such benefits and therefore view competency testing with favor.

As seen by several researchers (McKenna, 1979; Amarel, 1980; Bardon & Robinette, 1980), the fundamental conceptual fallacy of minimum competency testing, and its primary danger for classroom teachers, are that teachers are denied essential decision-making authority and are relegated to the position of "mechanics" in an educational machine. We have already seen that externally imposed tests strongly determine what is taught in the schools. Amarel (1980) cites the promulgation of highly prescriptive instructional programs often associated with competency testing and suggests that the tests may ultimately determine not only what teachers teach, but how they teach as well:

> Variously called diagnostic/prescriptive, individualized, or 'direct instruction' programs, [these highly prescriptive programs] have several features in common. Frequently, the prescription extends to the knowledge or skill domain to be taught (primarily basic skills), to the instructional sequence to be followed (usually linear), to the units of instruction (usually small and discrete), to the instructional process to be used (largely drill and practice), and to the specification of diagnostic or evaluative tools to be used (frequently paper-and-pencil tests). The professional autonomy of the teacher in a fully implemented program of this kind is diminished, since the teacher's role here calls for little active judgment or decision-making [p. 119].[3]

After noting the paradox between increasing demands to hold teachers accountable while, at the same time, limiting their discretion in selecting instructional goals, methods and standards, Amarel suggests that strict control of teachers' functions through competency testing and prescriptive instructional programs cannot help but lead to educational failure. The root cause of functional illiteracy, she notes, is not students' inability to master language mechanics, but their inability to comprehend—to "recover and reconstruct meaning that is imbedded in text." In Amarel's view, teachers who are limited in their choice of instructional purpose, method, and evaluation procedure cannot teach students to comprehend.

In a similar vein, Bardon and Robinette (1980) suggest that the loss of curricular freedom likely to result from competency testing is a personal

---

[3] Copyright 1980, American Educational Research Association, Washington, D.C.

danger for teachers and students. Such programs, they note, tend to make teaching less a profession and more a craft. Teachers are made to spend increasing amounts of time completing assigned tasks created by others and following prescribed routines. Bardon and Robinette predict increasing disillusionment on the part of teachers as a result, particularly if they are held responsible for educational outcomes while having little control of educational requirements, processes, and inputs.

In predicting the consequences of competency testing for students and teachers, we must be content, for the moment, with speculation rather than evidence. Logic tells us that minimum competency testing programs will produce neither better teaching nor greatly enhanced learning. Such outcomes are not legislated, so whether the negative outcomes suggested by some will actually come to pass cannot be predicted with certainty. If established standards of competence are low enough, hardly any students will fail the tests, dilution of the traditional high school curriculum will not be a threat, and teachers will be free to do what they have always done, and will be content in the knowledge that competency testing is a meaningless political ritual. Conversely, adoption of high standards, even for performance in such narrow curriculum areas as language mechanics and computation, could lead to wholesale failure of students, erosion of the decision-making authority of teachers, and the demise of the traditional K-12 curriculum.

Societal Consequences of Competency Testing

With an abundance of performance history as a guide, it is safe to predict that the societal effects of competency testing will not be uniform with respect to race and social class. The children of the poor will fail competency tests in far higher proportions than will the children of the rich. Coleman *et al.* (1966) found that the average achievement test scores of black children were about one standard deviation lower than the average achievement test scores of white children in both reading and mathematics. The early experience of states with uniform competency test standards is consistent with this reported disparity. For example, in the first operational administration of the North Carolina High School Competency Tests, 24.7% of black eleventh graders failed the reading test, while only 3.8% of white eleventh graders failed. Corresponding failure rates on the mathematics test were 33.8% for black students and 6.6% for white students.

In an analysis of the implications of such performance differences between black and white students, Eckland (1980) examined score distributions on tests administered in conjunction with the National Longitudinal Study (NLS) of the High School Class of 1972. Under the sponsorship of the National Center for Educational Statistics, the Educational Testing Service administered basic skills tests (similar in content to many minimum competency achievement tests presently in use) to a representative sample of about 18,000 high school seniors selected from over 1000 high schools throughout the United States. On a mathematics test consisting of computation items and questions requiring application of basic computation skills to simple word problems, 13.3% of the white students tested scored in the lowest 20% of the overall score distribution, while nearly half of the black students tested (49.8%) earned scores in the lowest 20% of the distribution. On a test composed of reading comprehension and vocabulary items (similar to the minimum competency achievement tests in reading used by many states), 12.8% of the white students scored in the lowest 20% of the overall distribution, while 50.9% of the black students scored in the lowest 20% of the distribution.

In a fascinating analysis of the importance of basic skills test performance to the economic survival of black and white students, Eckland (1980) examined the employment rates and income levels of students who scored in various portions of the overall distribution on the NLS tests. Among white students who did not enter college, he found that rates of unemployment shortly after high school graduation were virtually unrelated to performance on the NLS tests in reading and mathematics. Among black students who did not enter college, the rate of unemployment shortly after high school graduation was about twice as high for those who scored in the lowest half of the overall distribution on the math test as it was for those who scored in the upper half of the distribution. However, reading test performance bore only a slight relationship to unemployment for black high school graduates. Performance on the NLS test was found to be virtually unrelated to rates of part-time employment for both black and white students. Eckland also found that among high school graduates who gained full-time employment shortly after graduation, weekly income was virtually unrelated to performance on the NLS tests. About half the graduates (black and white) who were employed at least 35 hours per week earned less than $100 per week in 1972 regardless of performance on the NLS tests in reading and mathematics.

What should we make of these statistics? First, the economic certification value of a high school diploma is clear. In 1978, the unemployment rate of labor force members in the age range 16–21 who did not have a high school diploma was more than twice as high as the unemployment rate of high school graduates in the labor force who were in the same age range (U.S. Bureau of the Census, 1979, p. 148). Second, let us assume that states were to set passing scores on their competency tests at a level that failed high school seniors scoring in the lowest 20% of the score distribution. If a state's test really measured "minimal competence," such a standard would not appear to be overly stringent. With this passing standard, the proportion of black students denied a high school diploma would be four times as large as the corresponding proportion of white students. Thus, black students, to a far greater degree than white students, would be denied a level of certification that is demonstrably important to their economic well-being. This standard-setting would be done in the absence of evidence that the educational basis for withholding high school diplomas bears any clear relationship to employment and income potential.

Peng and Jaffe (1978) reported the results of a 4-year follow-up study of high school graduates who did not go to college based on NLS data. Among their findings on the relationship between test performance and economic well-being were (a) 4 years after high school graduation, 4.5% of those who scored in the top quarter of the NLS test score distributions were unemployed and seeking employment; (b) the corresponding percentage for those who scored in the lowest quarter of the test score distributions was 6.8; (c) the average weekly earnings of those who scored in the top quarter of the overall distributions was $175; and (d) the corresponding figure for those who scored in the lowest quarter of the distributions was $173 per week. On the basis of these data, Eckland (1980) concludes that "the relationship between test scores and economic success or failure (or more precisely, the almost total lack of such a relationship) has remained unchanged for the class of 1972 with the passage of four years [p. 131]."

It is not uncommon to hear the claim that minimum competency tests measure "survival skills." If economic survival is the referent of such claims, they are clearly unfounded. If helping students to gain employment and earn a living wage is one objective of public schooling, available evidence suggests that we jeopardize that goal to a far greater extent through failure to certify students than through failure to educate them. And one effect of minimum competency testing will surely be to exacerbate the current economic disparity between black and white youth.

All of the data considered thus far in examining the economic impact of competency testing for black and white students have been based on those who did not go to college. Eckland (1980) also examined the likely effects of competency testing on the college attendance rates of blacks and whites. As might be expected, the effects vary markedly, depending on the placement of passing scores. Among white students in the NLS sample 54% enrolled in college at some time during the 4 years following high school graduation. The corresponding figure for black students was 48%. Using the performance distributions for black and white students on the NLS tests, Eckland found that the college-going rate of white students would be virtually unaffected if students scoring in the lowest 10% or 20% of the distributions were denied high school diplomas. For black students, however, college-going rates would be affected dramatically by such policies. If students in the lowest 10% of the score distributions were denied high school diplomas, 10% fewer black students would be able to enter college. The difference in college-going rates between blacks and whites would increase from 6% to 16%. Setting competency test-passing scores at the twentieth percentile of the overall score distributions would reduce the white college-going rate from 54% to 52%—a negligible reduction. However, it would reduce the black college-going rate from 48% to 31%. The difference between white and black college attendance rates would thus be three and a half times as great as was found for the high school class of 1972. Since the economic value of college attendance is well known, these data further substantiate the disproportionate racial impact of competency testing.

CODA

The problems that inspired minimum competency testing programs in a majority of states and hundreds of school systems throughout the nation are real and undeniable. Tens of thousands of high school graduates read poorly, find the simplest computations beyond their meager abilities, and lack the skills needed to frame a coherent written paragraph. The educational meaning of a high school diploma is suspect. Taxpayers, state boards of education, state legislatures, and even high school students themselves, have rightful cause for concern.

Although the relationship between economic survival and basic school skills is questionable, the value of functional literacy in an increasingly technological society is obvious. Those who decry an overemphasis on basic skills in the school curriculum question sufficiency, not necessity. Fluency in the manipulation of symbols, be they letters, numbers or words, is essential to comprehending the ideas expressed in the language of those symbols. Is symbolic fluency essential to survival? Probably not. One could more readily argue that the schools offer nothing that is essential to survival, save a certificate that society endows with far more meaning than it rightfully deserves.

Is minimum competency achievement testing an appropriate remedy? Politically, perhaps it is, at least in the short term. Educationally, however, its worth is extremely doubtful, for both the short term and the long term. The mechanics of reading, writing, and computation are appropriate subjects for the elementary school, not for the high school. To insist, through edict or legislation, that high schools engage in an elementary-school level task will surely lead to inefficiency, and, most likely, to ineffectiveness. At best, the quality of American education will suffer. At worst, the concept of free, public education that covers 12 cumulative, enriching years of instruction may be destroyed. In the short term, the children of the poor and of the ethnic and racial minorities—those most likely to be denied high school diplomas—will suffer most. In the long term, all of us may lose.

REFERENCES

Airasian, P., Pedulla, J., and Madaus, G. *Policy issues in minimal competency testing and a comparison of implementation models.* Boston: Heuristics, Inc., 1978.

Amarel, M. Comments on H. Broudy's "Impact of minimum competency testing on curriculum." In R. M. Jaeger and C. K. Tittle (Eds.), *Minimum competency achievement testing: Motives, models, measures, and consequences.* Berkeley, California: McCutchan, 1980, 118–121.

American Friends Service Committee. *A citizen's introduction to minimum competency testing programs for students.* Columbia, South Carolina: Southeastern Public Education Program, 1978.

Baratz, J. C. Policy implications of minimum competency testing. In R. M. Jaeger and C. K. Tittle (Eds.), *Minimum competency achievement testing: Motives, models, measures, and consequences.* Berkeley, California: McCutchan, 1980, 49–68.

Bardon, J. and Robinette, C. Minimum competency testing of pupils: Psychological implications for teachers. In R. M. Jaeger and C. K. Tittle (Eds.), *Minimum competency achievement testing: Motives, models, measures, and consequences.* Berkeley, California: McCutchan, 1980, 155–171.

Beard, J. G. Minimum competency testing: A proponent's view. *Educational Horizons, 58,* Fall, 1979, 9–13.

Blau, T. H. Minimum competency testing: Psychological implications for students. In R. M. Jaeger and C. K. Tittle (Eds.), *Minimum competency achievement testing: Motives, models, measures, and consequences.* Berkeley, California: McCutchan, 1980, 172–181.

Broudy, H. S. Impact of minimum competency testing on curriculum. In R. M. Jaeger and C. K. Tittle (Eds.), *Minimum competency achievement testing: Motives, models, measures, and consequences.* Berkeley, California: McCutchan, 1980, 108–117.

Broudy, H. S., Smith, B. O., and Burnett, J. R. *Democracy and excellence in American secondary education.* Chicago: Rand McNally, 1964.

Callahan, R. *Education and the cult of efficiency.* Chicago: University of Chicago Press, 1962.

Cohen, D. and Haney, W. Minimums, competency testing, and social policy. In R. M. Jaeger and C. K. Tittle (Eds.), *Minimum competency achievement testing: Motives, models, measures, and consequences.* Berkeley, California: McCutchan, 1980, 5–22.

Coleman, J. S., Campbell, E. Q., Hobson, C. J., McPartland, J., Mood, A. M., Weinfeld, F. D., and York, R. L. *Equality of educational opportunity.* Washington, D.C.: U.S. Government Printing Office, 1966.

Donahue vs. Copaigue Union School Free School District. 407 N.Y.S. 2d 874 (Supreme Court, App. Div., Sec. Dept. 1978) Affirming, Index 77-1128, Opinion 8/31/78.

Eckland, B. K. Sociodemographic implications of minimum competency testing. In R. M. Jaeger and C. K. Tittle (Eds.), *Minimum competency achievement testing: Motives, models, measures, and consequences.* Berkeley, California: McCutchan, 1980, 124–135.

Fremer, J. In response to Gene Glass. *Phi Delta Kappan,* 1978, *59,* 605–606, 625.

Gallagher, J. J. Setting educational standards for minimum competency: A case study. In R. M. Jaeger and C. K. Tittle (eds.) *Minimum competency achievement testing: Motives, models, measures, and consequences.* Berkeley, California: McCutchan, 1980, 239–257.

General Assembly of the State of North Carolina. Article 39A, Section 115-320.6, "Purpose," 1977.

Gentry, C. G. Will the real advantage of CBE please stand up? *Educational Technology,* 1976, *16,* 13–15.

Gorth, W. P., and Perkins, M. R. *A study of minimum competency testing programs.* Final program development resource document. Amherst, Massachusetts: National Evaluation Systems, Inc., December, 1979a.

Gorth, W. P., and Perkins, M. R. *A study of minimum competency testing programs.* Final typology report. Amherst, Massachusetts: National Evaluation Systems, Inc., 1979b.

Hart, G. K. The California pupil proficiency law as viewed by its author. *Phi Delta Kappan,* 1978, *59,* 592–595.

Hentoff, N. *Minimum competence: Part One: Program no. 123, options in education* (transcript). Washington, D.C.: National Public Radio, June 5, 1978.

Herron, M. D. Graduation requirements in the state of Oregon: A case study. In R. M. Jaeger and C. K. Tittle (Eds.), *Minimum competency achievement testing: Motives, models, measures, and consequences.* Berkeley, California: McCutchan, 1980, 239–257.

Jaeger, R., Cole, J., Irwin, D., and Pratto, D. *An iterative structured judgment process for setting passing scores on competency tests applied to the North Carolina high school competency tests in reading and mathematics.* Greensboro, North Carolina: Center for Educational Research and Evaluation, University of North Carolina at Greensboro, 1980.

Jencks, C., Smith, M., Acland, H., Bane, M., Cohen, D., Gintis, H., Heyns, B., and Michelson, S. *Inequality: A reassessment of the effect of family and schooling in America.* New York: Basic Books, 1972.

Lindquist, E. F. Considerations in objective test construction. In E. F. Lindquist (Ed.), *Educational Measurement.* Washington, D.C.: American Council on Education, 1951.

Lott, W. *Personnel and guidance association newsletter.* Albany, New York: New York State Personnel and Guidance Association, Spring, 1977.

McKenna, B. Minimal competency testing: The need for a broader context. *Educational Horizons,* 1979, *58,* 14–19.

Madaus, G., and Airasian, P. *Measurement issues and consequences associated with minimal competency testing.* National Consortium on Testing, May 1978.

Miller, B.S. (Ed.), *Minimum competency testing: A report of four regional conferences.* St. Louis, Missouri: CEMREL, 1978.

National School Boards Association. *Minimum competency.* Research report. Washington, D.C., 1978.

Oregon Administrative Regulations, 581-22-200, 581-22-228, 581-22-231, 581-22-236.

Peng, S. and Jaffe, J. *A study of highly able students who did not go to college.* Technical report. Research Triangle Park, North Carolina: Research Triangle Institute, September, 1978.

Perrone, V. Competency testing: A social and historical perspective. *Educational Horizons,* 1979, *58,* 3–8.

Peter W. vs. San Francisco Unified School District. 1976. 60 Cal. App. 3d 814, 131 Cal. Rptr. 854 (Ct. App. 1976).

Reilly, W. Who benefits? Competency testing. *Compact 12,* 1978, 7.

Ryan, W. *Blaming the victim.* New York: Pantheon Books, 1971.

Tyler, R. W. The minimal competency movement: Origin, implications, potential and dangers. *National Elementary Principal,* January 1979.

U. S. Bureau of the Census. *Statistical Abstract of the United States* (100th ed.). Washington, D.C., 1979.

Chapter 11

The Implications for Society

Gilbert R. Austin
Herbert Garber

Studies of student performance using "then and now" data (to use Farr's words) require that we first describe as adequately as we can what it was like "then" and what it is like "now" so that gains or declines can be put in appropriate contexts. Almost all of the authors have referred to James Coleman's *Equality of Educational Opportunity* (1966). The most widely quoted statement from that book is "only a small part of achievement variation is due to school factors. More variations are associated with the individual's background than with any other measure, but most test score variation is not associated with any measure in the survey [p. 298]." A similar study done in England, "Children and Their Primary Schools" (commonly referred to as the Plowden Report), says "children's work and behavior in school are profoundly influenced by their home circumstances [Plowden committee, 1967, p. 347]."

Coleman and the Plowden Report are talking about comparisons between children who are *attending* school. Both Coleman and the Plowden Report acknowledge that schools make an important and significant contribution to children's lives and to equalizing their life chances. At the most fundamental level, this point is made by comparing the effect of schooling with that of no schooling. Unfortunately, there are few studies in the literature on this topic. One such study, Green *et al.* (1964), was conducted in Prince Edward County, Virginia. This county closed its public schools from 1959 to 1961 rather than comply with a court order to integrate them;

247

Copyright © 1982 by Academic Press, Inc.
All rights of reproduction in any form reserved.
ISBN 0-12-068580-9

thus, some children in the county did not go to school during this period. Comparisons of the children who stayed in the county but did not go to school with the children who went to private school or who left the county to attend school elsewhere showed that the children who did not attend school not only failed to progress but, in many cases, seriously regressed in their intellectual abilities and in school-related achievement.

In their chapter on education in Great Britain and Ireland, Kellaghan and Madaus make the same point about children whose education was interrupted by World War II. Their data indicate that not only does school help all children, of whatever ability, to make progress, but its absence causes them to regress in their abilities. Making this point from a different perspective, Jencks *et al.* (1972) states that "elementary schooling is helpful for middle-class children and crucial for lower-class children [p. 89]." Two studies (Schmidt, 1966, done in South Africa, and Vernon, 1969 done in Nigeria and Senegal) showed similar findings when the researchers compared the performance of children who had the benefit of schooling with those who did not. The International Educational Achievement (IEA) studies (Madaus *et al.*, 1980) indicate that "the proportion of variance in achievement accounted for by school factors (as compared to the proportion accounted for by out-of-school factors) was much greater in developing countries than in industrialized developed countries [p. 36]".

We have been very successful at educating our children; perhaps this is part of the problem. In 1800, only 5 out of every 100 children in the United States graduated from high school. In 1981, it is estimated that 75% of the 18-year-old males and 80% of the females will graduate from high school (NCES, 1974). This is indeed an impressive indicator of progress in the last 180 years.

We have attempted to indicate in the previous paragraphs that we are trying to view educational test score gains and losses within the context of a *changing* educational and social milieu, whereas vocal critics view the context as unchanging.

The editors asked a straightforward question at the beginning of the book: "In what ways have school test scores been changing during recent years, and what factors account for those changes?" In this book, the authors' findings indicate that school test score changes occur in both directions, up as well as down. They vary for achievement tests across subject matter; geographic region; ethnic group and social class; age; sex; and under conditions of war; social unrest; peace; economic decline or prosperity; changing family patterns; and changing political policies. College admissions test scores, alone, have recently been unvarying in their sustained decline.

The greatest difficulties lie in identifying specific causes for specific changes, in collecting appropriate information for proper analyses of the problem, and in answering one crucial question which was only partly answered: Do the schools determine what students shall learn and how well they shall learn it?

The goal of this book was to present some facts, with judicious interpretations, that might be useful in resolving existing conflict and controversy. The issue of changing test scores has become controversial because, for whatever reasons, the public's attention has been focused only on those scores which declined. A more balanced presentation documents the rises as well as the stabilities and declines.

As Bruce Eckland pointed out, the Wirtz Commission cautioned that whatever is wrong with the cognitive development of American youth will require "collaboration of teachers, students, parents and *the broader community* to correct [stress added]." This leads to an observation which too frequently is ignored. Every society expects its schools to reflect and teach only those cognitive and affective learnings that it values; it would be folly to expect the school to develop aims and goals at variance with those of the community it serves. It follows, therefore, that one could reasonably ask if it is possible that changing test scores are a fairly accurate, though lagging, measure of social aspirations and conditions?

The authors of this book independently and fairly consistently brought forth several themes. They found that test scores have both increased and decreased; that student populations have changed with fewer dropouts leading to lower average ability levels in grades 8–12; that curricula and tests have changed, and not necessarily in compatible ways; and that it is nearly impossible to collect dependable test data within a content field on a nationwide level.

Changing national school policy, as in racial integration through deliberate busing, appears to have no clear effect on school achievement, self-esteem, race relations or success in higher education. Family pattern changes including increased numbers of working mothers (in 1981, for the first time, females made up more than 50% of the United States work force), confluence theory on birth order, rising divorce rates, and one-parent families all probably account for some score declines, but these factors are difficult to measure systematically.

The authors refer also to changes promulgated outside the schools, changes which have been affecting school practices and student attainments. The Lyndon Johnson presidency launched an effort to bring this nation to a condition of educational and economic equity. The schools chose to invest most of these new resources primarily on the early years of education (preschool through elementary), reflecting the writing of Hunt (1961) and Bloom (1964). These programs were aimed at children from low-income families via Head Start, Follow Through, and The Elementary and Secondary Education Act (ESEA) Titles I and III. For older students, the Job Corps, Vista, and the Trio program were created to enhance employability through vocational training service and college education.

Test scores at the lower elementary grades have not declined in many areas but have actually been improving. It is not possible to assert from our data which causes can be credited. However, it seems reasonable to point to the early childhood programs just identified as likely contributors to these positive outcomes.

One of the important points that has come out of the intensive investigation of preschools and early elementary schools in the last 15 years is the idea of developmental continuity, which concerns the transition from preschool to primary school.

In England, Dr. A. Halsey of Nuffield College, Oxford noted in one of three general conclusions about preschool and child development that, if properly understood, preschooling is a point at which the networks of family and formal education can most easily be linked. Parental concern and involvement in preschool have always been greater than in later education. Mothers and preschool teachers in all countries commonly meet and discuss the child's needs and problems. Some parents have recently been demanding even greater involvement. They want to have more voice in the choice of curriculum and in the planning and organization of the preschool.

One clear finding emerges from a review of the importance of preschool and elementary education and the orientation it should take. That is, whatever goals are desired in early childhood education (ECE), the achievement of them will be more certain if the program reflects the following characteristics (Austin, 1976):

1. Its values and practices are in consonance with those of the parents, teachers, and community.
2. Its organizational characteristics reflect a high commitment to the project on the part of the preschool principal and other school administrators.
3. It provides a staff training for teachers that is oriented toward the accomplishment of the program's goals.
4. It provides for parental involvement.
5. It reflects careful planning and frequent evaluation and modification of the program in light of what evaluation reveals about the successes and failures of the program [p. 74.].

Wargo (1977), in a paper entitled "Those Elusive Components that Contribute to Success of Compensatory Education Projects," identifies a surprisingly similar list of characteristics for elementary school. Wargo maintains that

1. Academic objectives be clearly stated
2. There be active parental involvement, particularly as motivators
3. There be individual attention for pupils' learning problems
4. There be a high intensity of treatment

Recent findings by Schweinhart and Weikart (1980), in their report, which is one of the first to follow preschoolers through high school, indicate that preschool programs for the disadvantaged pay off in better school achievement, lower discipline problems and a more positive attitude about school in general. The authors say that there is also a pay-off in dollars from preschool programs: They suggest that for a $6000 investment parents and society get back a $15,000 savings. The same point is made by Zigler *et al.* (1979) in a recent book entitled "Project Head Start, A Legacy of the War on Poverty."

As many of the authors have indicated in this book, almost all of the evidence presented indicates that children up through grades 3 or 4 are, in general, obtaining higher achievement test scores now than they formerly did. The reasons for this are not clear or easily evident, nor is there any ability to state causal factors with a great degree of assurance. What does seem to have had an effect is the fact that since 1965, we have spent a great deal more time, money, energy, and professional talent on researching the problems of educating children between the ages of 4 and 8.

Television has been, paradoxically, cited as both villain and hero. Educational programming in the ECE area (e.g., *Sesame Street, The Electric Company,* etc.) seems to have contributed to test score gains for elementary school children, while spending time with television rather than with reading and writing is cited by more than one author as a partial explanation for declining scores for young people over the age of 9.

This research seems to suggest that the effects of excellent educational television programming are clearly beneficial. At the same time, serious questions must be raised about television for the older child.

By the time a child graduates from high school, he has watched up to the equivalent of fifteen thousand hours of television, which is approximately equivalent to the number of hours he has spent in school. The authors suggest that a great deal of this viewing has been done at the expense of needed independent reading and completion of homework assignments. In general, television is an important distractor from school-oriented achievement. This suggests that we should be looking very seriously at the kinds of television programs being offered and at their educational content for older elementary, junior, and senior high school students. Perhaps we should try to create a program to build upon the positive effects of the early education efforts.

National policy, concerned with equality of educational (and economic) opportunity, made staying in school and going to college a desirable choice for great numbers of low-income students who also had a generally lower academic ability. However, although Eckland points to this phenomenon as a definite contributor to declining SAT scores for high school juniors and seniors, Flanagan presents high school achievement test data which suggest strongly that abilities among high school students remained stable on average but with notable changes in the pattern of scores.

Several of the authors discussed the question of teacher competency and the possible influence of the same changing social policy on the quality of those who enter the teaching profession as well as the pupils who attend school. Although the evidence for a direct causal relationship between teacher competence and student learning is lacking, it seems logical to expect that a policy directed toward attracting more competent school personnel into the classroom and principal's office should be a means to a desirable end. Coleman *et al.* (1966) indicate that teachers' verbal skills have a strong effect on pupil achievement which first shows itself at the sixth grade (p. 318). Eckland, citing research by Weaver (1979), suggests that new recruits being attracted to the teaching profession may have lower academic potential than their earlier cohorts.

What about the courses of study that upper level students take? There is reasonably good evidence that high school students are, with increasing frequency, choosing to avoid the hard sciences (i.e., advanced mathematics courses; advanced physics, biology and chemistry courses) and are substituting courses in the humanities, which may or may not be more intellectually demanding. Certain achievement and aptitude test score declines can be traced to the fact that students simply have not taken the courses that would have given them the ability to answer the questions posed.

For a variety of economic reasons, increasing numbers of students are staying in school for longer and longer periods of time. Greater numbers of students mean that the general level of achievement in the upper school grades has dropped as compared to the attainments of the more selective student population of some 10 or 20 years ago. As Husen (1967) of Sweden has said about making school open and available to a larger and larger portion of the population, "more must mean less"; this is to say that when children from social classes that formerly were excluded begin to attend school, average test scores can be expected to decline since the new subpopulation may not have the needed social–cultural background for school success. It is interesting that a country with no significant racial minority

(Sweden) should experience a test score decline similar to the one in America, where class and race are strongly associated.

Flanagan's Project Talent findings and Forbes' National Assessment of Educational Progress score trends both suggest that even though students of lower ability have become a larger component in the high school population, scores on tests which measure outcomes of certain school-related activities have remained stable or even increased. Farr and Fay substantiate this same notion in discussing and describing reading trend data. They point out, too, that when scores are compared between two time periods (about 10 years apart), the average age of the more recent group is usually significantly lower. Adjustment for this age difference almost invariably shows today's student, *at any grade level*, to be outperforming the student from the 1950s and 1960s.

Another important theme which recurs in the chapters of this book is concerned with the technical problems associated with the testing instruments themselves. We are shown, for instance, that people can be made to appear more or less literate by the simple device of moving cut-off scores down or up. Achievements in mathematics, science, reading, and writing (and their interpretations) are, to a significant extent, functions of test item quality, required cognitive ability, scoring methods, and choice of score transformation (e.g., grade equivalents or stanines). No firm evidence emerged, however, to allow us to blame test technology for any undesirable testing outcomes in general.

Related to such psychometric issues is national educational policy. Kellaghan and Madaus point out (in comparing norm-referenced standardized tests with subject-specific "public examinations" in Britain and Ireland) that norm-referenced tests are sensitive to a general trait which correlates more with home background variables than with specific school variables. This seems to confirm the Coleman (1966) and Jencks (1972) findings with respect to national tests. On the other hand, subject-specific tests are more school-sensitive since their content is dominated by exercises for which pupils from outside the school setting are not likely to be prepared. The implication is that falling nationwide test scores (norm-referenced) tell us very little that is useful about the school as an educational institution but may tell much about other community (national) educational factors. These nonschool factors make their influence noticeable on standardized test scores but probably fail to show the school-specific gains that the same students have made (see Flanagan's Project Talent score gains).

Finally, one theme did come forth that needs particular emphasis. Jaeger warned us most forcefully and Harris and Amprey urged us to consider that certain positive steps be taken to deal with the potential for increased differentiation between the socioeconomic classes. Policymakers would be well-advised to move cautiously when considering mandating competency achievement testing for pupil promotion or graduation. It seems that the American proclivity for action in the face of a problem might not only fail to solve the original problem but could create even worse ones. The feelings of despair endured by so many of our citizens over having the doors to economic opportunity slammed against them could easily become aggravated from hastily adopting testing procedures which will not increase "competence" so much as deny "credentials."

This book has produced few hard answers to the original questions posed. However, it has indicated that the controversy, like most controversies, is based more on sensationalism than on reasoning. Nevertheless, people make decisions on the basis of what they regard as the truth. Hopefully, this book has shown that American schooling and test scores are intimately connected to the society they serve. Only a complex network model could approach conclusive answers to the variety of questions raised. One fact has emerged: The wages of American society are the scholastic achievements of its children. The reader must decide how to evaluate both.

REFERENCES

Austin, G. *Early childhood education: An international perspective.* New York: Academic Press, 1976.

Bloom, B. *Stability and change in human characteristics.* New York: John Wiley and Sons, 1964.

Coleman, J. S., *et al. Equality of educational opportunity.* Washington, D.C.: United States Government Printing Office, 1966.

Green, R. L., Hofman, L. T., Morse, R. J., Hayes, M. E., and Morgan, R. F. *The educational status of children in a district without public schools.* Cooperative Research Project No. 2321. Office of Education, U.S. Department of Health, Education, and Welfare. Lansing, Michigan: Michigan State University, College of Education, 1964.

Halsey, A. H. *Educational priority: EPA problems and policies* (Vol. 1). London: HMSO, October, 1972.

Hunt, J. *Intelligence and experience.* New York: Ronald Press, 1961.

Husen, T. *International project for the evaluation of educational achievement.* International Study of Achievement in Mathematics: A Comparison of Twelve Countries. Vol. I and II. Stockholm: Almqvist and Wiksell, and New York: Wiley, 1967.

Jencks, C., Smith, M., Acland, H., Bane, M. J., Cohen, D., Gintis, H., Heyns, B., and Michelson, S. *Inequality: A reassessment of the effect of family and schooling in America.* New York: Basic Books, 1972.

Madaus, G., Airasian, P., and Kellaghan, T. *School effectiveness: A reassessment of the evidence.* New York: McGraw-Hill, 1980.

National Center for Educational Statistics. *Projections of educational statistics to 1982-83.* 1973 edition. Office of Education, U.S. Department of HEW. Washington, D.C., 1974.

Plowden Report. *Children and their primary schools.* Central Advisory Council for Education, HMSO 1967, Vol. I and II.

Schmidt, W. Socioeconomic status, schooling, intelligence, and scholastic progress in a community in which education is not yet compulsory. *Paedogogica Europa,* 1966, *2,* 275–286.

Schweinhart, L. J., and Weikart, D. P. *Young children grow up: The effects of the Perry pre-school program on youths through age fifteen.* Ypsilanti, Mich. High School Press, 1980.

Vernon, P. *Intelligence and cultural environment.* London: Methuen, 1969.

Wargo, M. *Those elusive components that contribute to the success of compensatory education projects.* Paper presented at the 1977 Annual Meeting of the American Educational Research Association, April 4–8. New York City.

Weaver, W. T. In search of quality: The need for talent in teaching. *Phi Delta Kappan,* September 1979, 29–46.

Zigler, E., and Valentine, J. *Project Head Start: A legacy of the war on poverty.* New York: Free Press, 1979.

Index[1]

Compiled by Hans H. Wellisch

[1]Entries are arranged word by word. Numerals precede letters. Abbreviations are filed as words, and exactly as spelled (nor as pronounced). References to tables are marked with an asterisk (*). References to footnotes are marked with the letter n. Page references in *italics* refer to bibliographies.